The Economics of the Petrochemical Industry

Marwan Fayad
Gulf Organization for Industrial Consulting, Doha, Qatar

and

Homa Motamen
Imperial College, London University

St. Martin's Press, New York

To Our Mothers

St. Martin's Press, Inc., 175 Fifth Avenue, New York, NY 10010
Printed in Great Britain
First published in the United States of America in 1986

ISBN 0-312-23444-9

Library of Congress Cataloging in Publication Data

Fayad, Marwan.
The economics of the petrochemical industry.

Bibliography: p.
Includes index.
1. Petroleum chemicals industry. 2. Petroleum chemicals industry–Technological innovations.
I. Motamen, Homa. II. Title.
HD9579.C32F39 1985 338.4'7661804 85-2099
ISBN 0-312-23444-9

CONTENTS

LIST OF TABLES

LIST OF FIGURES

PREFACE

This book is concerned with the study of the economic and technological factors that reshaped the petrochemical industry at the various stages of its development. Its main purpose is to shed some light on those factors, putting in proper perspective the recent developments in the industry on the world-wide scene, and to analyse the changing roles of the main petrochemical producers, so as to add to the understanding of the dynamic structure of the industry and to allow better planning of its future course of developments, taking into consideration the growth potential of its markets.

The book shows that the diffusion of the process technology by the imitator producers, together with the successive oil price rises of the 1970s, which effectively changed the production cost structure of petrochemical products in favour of the feedstock component, have contributed to the increased maturity of the industry, resulting in a slow, but steady, change in the roles of the main producers. Where the traditional major chemical companies, with their strong R & D base and technological know-how, direct their future investments into the production of sophisticated chemicals with higher value added.

On the other hand, the major oil companies with their hold on oil supplies have been increasing their participation in petrochemical production, a process that will continue in the future, when major oil companies would account for a larger share of new petrochemical capacities for bulk products, at the expense of the major chemical companies. However, despite the increased maturity of the industry, there has been an increased concentration of production capacities and market shares in the hands of the largest producers in industrialised countries. Since the pressures of cost reduction efforts require the exploitation of economics of scale and correspondingly huge financial commitments.

Also, the oil producing countries with their access to cheap energy sources have the opportunity to participate more actively in petrochemical production and the export of bulk petrochemicals.

This book also establishes that industrial production serves as a better indicator of the demands for petrochemical products than Gross Domestic Product; and, now that these products are mature, detailed input–output models would give reliable forecasts of future demands for these products.

Marwan Fayad,
Homa Motamen
May 1985

INTRODUCTION

The petrochemical industry was one of the fastest growing and among the most profitable industries from the 1950s to the 1970s. However, in the late 1970s continuing decline in the industry's growth rates, accompanied by widespread overcapacity and plant closures in most of the traditional industrialised countries, resulted in great controversy about the future of the industry. On the other hand, there were increased additions to capacity and plant building throughout the world, notably in the oil-producing countries, but also increased uncertainty about the availability of feedstock supplies. In the light of the much publicised 'energy problem', efforts are being made to develop non-oil feedstocks (SNG, methanol), but the industry will continue to be oil- and gas-based in the foreseeable future. Also, considerable attention has recently focused on the development of new integrated petrochemical systems and complexes with more flexible processes which employ alternative routes that avoid use of scarce or costly intermediate products and raw materials and simplify production. Extensive work has been carried out on developing these systems by a research group, headed by Professor D. Rudd (see Bibliography) at the University of Wisconsin — Madison. Although these issues have not been addressed in this book, their importance and the impact of the new sytems on the future structure of the petrochemical industry should not be underestimated.

This book, however, examines developments in the petrochemical industry at various stages, from its early years at the beginning of this century, through the heyday of the 1960s, up to the present transformation, analysing the economic and technological factors that led to the rapid growth of the industry and its dynamic and changing structure. Particular attention is paid to the changing structure of production costs, resulting from higher oil prices and the increased maturity of the process technology, and the impact of these changes on the roles of main petrochemical producers in industrialised countries and the implications for oil-producing countries. The future growth prospects for the industry are also discussed.

To understand a certain phenomenon and to explain concurrent developments they should wherever possible be related to a sound theoretical framework. This allows for a proper advance planning and anticipation of future developments in similar or alternative circumstances. We discuss the developing trends in the petrochemical industry in the context of a theoretical framework to elucidate the industry's dynamic structure. It is hoped that this will give some credibility to the conclusions arrived at in this book, give rise to better ways of looking for solutions to problems that may yet occur, and so steer future developments in the industry to a safer course.

Following this introductory chapter, the main characteristics of the petrochemical industry, the market forces that led to the rapid expansion in the demand for petrochemical products and the dynamic and changing structure of the industry are discussed in Chapter 1.

The ensuing discussion shows that the availability of cheap energy resources and the advances made in polymer science and organic chemistry, and in the process and chemical engineering technology, were responsible for the rapid growth of the industry on the supply side. While the expansion of markets, the rising industrial output and competitive prices of petrochemical products were responsible for the growth on the demand side.

The rapid technological changes and advances, in the industry, are attributed to the 'clustering' of innovations over various periods during the development stages of the industry (Freeman, 1974), and the dynamics of the industry are explained within the framework of a Schumpeterian-type growth cycle (Freeman, 1982), where the role of 'innovator' and 'imitator' producers in expanding the output of the industry and in increasing its production capacities are analysed. These changing roles of petrochemical producers are studied further in Chapter 2 and Chapter 3.

The role of the increased activity of 'imitator' producers and the changing structure of production costs in pushing the industry into its 'mature' stage (Stobaugh, 1968) are also touched on in this chapter but dealt with in more detail in Chapter 2, and the international implications of this process are dealt with in Chapter 4.

The remaining sections of Chapter 1 deal with the supply and demand patterns of major petrochemical products and the effects of oil price rises on the competitivity of plastic materials and their future use in two potentially large markets, the car industry and the building industry.

In Chapter 2, the changing structure of production costs of petrochemical products, resulting from oil price rises and technological advances, and the implications of these changes for main petrochemical producers are discussed.

It is also shown that, despite rising capital investment costs of petrochemical plants, which have continued to increase above the rate of inflation since the early 1970s, the raw materials component continued to dominate the final production costs of petrochemical products; since two factors add to the significance of the raw materials component, the increased exploitation of the benefits of economies of scale, with the effect of reducing capital costs per unit of output, and the increase in oil prices at higher rates than capital investment costs. The oil price rises are shown to be most significant for basic petrochemical products (such as ethylene, ammonia, etc.) and bulk plastics materials (LDPE, PVC), where raw materials costs account for about 70–80 per cent of the total production cost.

As a result of the changing structure of production costs of petrochemicals, a process of restructuring the operations of the industry has been set in motion. The roles of its operators, the major chemical and oil companies, in recent developments in the West European and American scenes and their future respective roles in this process are also discussed in this chapter.

Another aspect, very characteristic of the petrochemical industry and dealt with in Chapter 2, is the oligopolistic control of the markets by leading firms and the increased concentration and integration of petrochemical production as the industry becomes more capital-intensive, with very large plants coming on stream in the markets of traditional producers of industrialised countries. So the increased maturity of the industry and the standardisation of the technology have been translated into an increased number of producers, but these remain very few due to the huge financial requirements for the modern petrochemical complexes.

Chapter 3 discusses the petrochemical industry in the United Kingdom and is basically divided into two main parts. The first part demonstrates the usefulness of input–output techniques in estimating the demands for major petrochemical products and in defining the intersectoral relationships that exist between the synthetic resins and plastics materials industry and its major markets. Also proposed is a methodology for forecasting future demands for petrochemical products within the framework of input–output models. Although the analysis has been devoted to the British petrochemical industry, it should be noted that the methodology of the analysis could also be applied to any other producing country.

The second part of Chapter 3 discusses the supply side of petrochemicals and the structure of the British petrochemical industry. Particular attention is devoted to recent developments that have been taking place in the United Kingdom, the main problems facing the industry and the general performance of the British industry relative to other major West European producers.

Chapter 4 discusses the implications of increased maturity of the petrochemical industry and the changing production cost structure of petrochemical products for increased petrochemical production in developing countries, where it is argued that, since raw materials costs account for most of the production costs of basic petrochemical products and bulk plastics materials, many developing countries could satisfy their petrochemical needs by installing medium or small plants; while major oil-producing countries, with their access to vast quantities of cheap energy resources, have the opportunity to upgrade the value of their hydrocarbon resources by producing petrochemicals for export markets.

We also look at the role of the chemical and oil companies in joint ventures with oil-producing developing countries and the main obstacles facing petrochemical exports from these countries to major markets of industrialised countries. The development of the petrochemical industry in newly industrialised and oil-producing developing countries and their patterns of production and consumption of major petrochemical products are also discussed in this chapter.

Finally, the concluding chapter summarises the main themes of this study by drawing on the conclusions of each chapter. Also, the dynamic forces that have reshaped the petrochemical industry are put in proper perspective and a few recommendations concerning the developments in petrochemical sectors of developing countries are presented.

1 CHARACTERISTICS AND MARKET FORCES

1.1. Introduction

The origins of the petrochemical industry can be traced back to the evolution of the organic chemical industry at the beginning of this century, with the production of a few thousand tons of synthetic dyestuffs in Germany. Petrochemical products are essential for the production of a diverse set of products such as plastics and resins, synthetic fibres and rubbers, solvents and paints, fertilisers, drugs explosives and many more. However, this book will concentrate on the study of basic petrochemical products and plastics materials, but reference will also be made to other types of products.

The production of plastics materials has grown enormously over the last twenty years. Its world-wide output in volume terms could approach that of steel by the end of the century. Plastics are used in all walks of life, from bank credit cards to plastic bags. They are used in hospitals and offices but more widely as household items. They also find applications in cars as well as the latest designs of fighter and outer-space aircraft.

As an introduction, the following pages will concentrate on the study of the development of the petrochemical industry from its early stages to the present day. The second section of this chapter analyses the trends in the production and consumption of the major plastics materials in the major producing regions of the United States, West Europe and Japan, and the future prospects for petrochemical production and consumption in each region. The applications of plastics materials in two of their large markets, the car industry and the building industry, and the potential for their increased use in the future in these two areas are also studied.

1.2. The main characteristics of the petrochemical industry

The petrochemical industry of today is characterised by the wide variety of its products and their end uses, the complexity of production, the alternative routes of production processes to final products and the flexibility in the choice of feedstocks. Petroleum-derived chemicals are numerous; they include over 90 per cent of organic chemicals and a significant amount of inorganics such as ammonia, ethylene, sulphur and carbon black. The large bulk of petrochemical products is consumed in the form of intermediates for the production of plastics, synthetic rubber and synthetic fibres, fertilisers, detergents and pesticides. The production of these materials has increased rapidly from about 3 million tonnes in 1950 to over 70 million tonnes by the mid-1970s, with plastics accounting for over half of this production. The chemical industry and particularly the petrochemical sector was one of the fastest growing industries during this period with outputs growing at 2.5 to 3 times the rate of growth of the Gross Domestic Product in Western Europe.

The industry's high growth rates, in the range of 15 per cent to over 20 per cent per annum, were a result of the combination of vast capital investments and R & D expenditures which resulted in rapid technological improvements on one hand, and the availability of cheap raw materials and low energy costs on the other, which allowed the new products to supply the increasingly growing markets at competitive prices, substituting many natural materials as well as finding new applications.

However, since the early 1970s the slowdown in the growth of the petrochemical industry was becoming evident even before the oil crisis of 1973 (Reining, 1982), which concealed the early signs of the industry's maturity at first but later helped in bringing it about. The lower growth rates, the spreading of overcapacity, particularly in Western Europe, and the sensitivity of production costs to factor inputs of raw materials and capital costs in the late 1970s are all signs of this maturity.

Another characteristic of the mature petrochemical industry is the small number of producers involved in petrochemical production, the bulk of the industry's output being consumed in the form of intermediate products by an equally small number of industries.

1.3. Technical innovation, demand and the growth cycle theory

It is appropriate at this stage to put in proper perspective the theoretical background that should provide a basis for understanding as well as explaining the dynamics and the driving forces behind developments in the petrochemical industry, whose full details are analysed briefly in the following sections of this book.

Technical innovation played a significant role in the development of this research-intensive industry. The various theories on the subject have been studied by Freeman, paying close attention to the electrical and the chemical industry, particularly the synthetic materials sector (Freeman, 1982). His careful examination of Schumpeter's theory of business cycles and industrial innovation confirms that the development of the plastics industry fits adequately into the successive stages of this theory (described briefly below).

Freeman attributes the main impetus behind the rapid growth of the plastics industry to a combination of two interdependent factors: an *inventive push*, resulting from exogenous/indigenous science and technology, based on a cluster of inventions in the early stages, and a *demand pull*, influenced by the growing markets and fluctuating economic activity. The two factors have a shifting relationship which changes continuously over time, as postulated by Schumpeter's Mark II model, shown in Figure 1.1, in which there is a strong feedback loop from successful innovations to increased R & D activities in a self-generating cycle where the market structure continues to grow and change with increased concentration in the hands of the largest firms, which establish a competitive lead through their inventive/innovative activities and increased investments.

However, the large profits achieved initially by the innovating firm would attract a *swarm* of imitators and improvers to join the original producer(s), who would be enjoying monopolistic (oligopolistic) profits in exploiting the new opportunities and markets created by the innovation (product or technology). The *swarming* process may take place slowly over a long period, but once this process starts it has powerful multiplier effects in generating additional demands on the economy for materials and products, machinery, services facilities and labour.

As the industry matures, profits will be squeezed under pressure

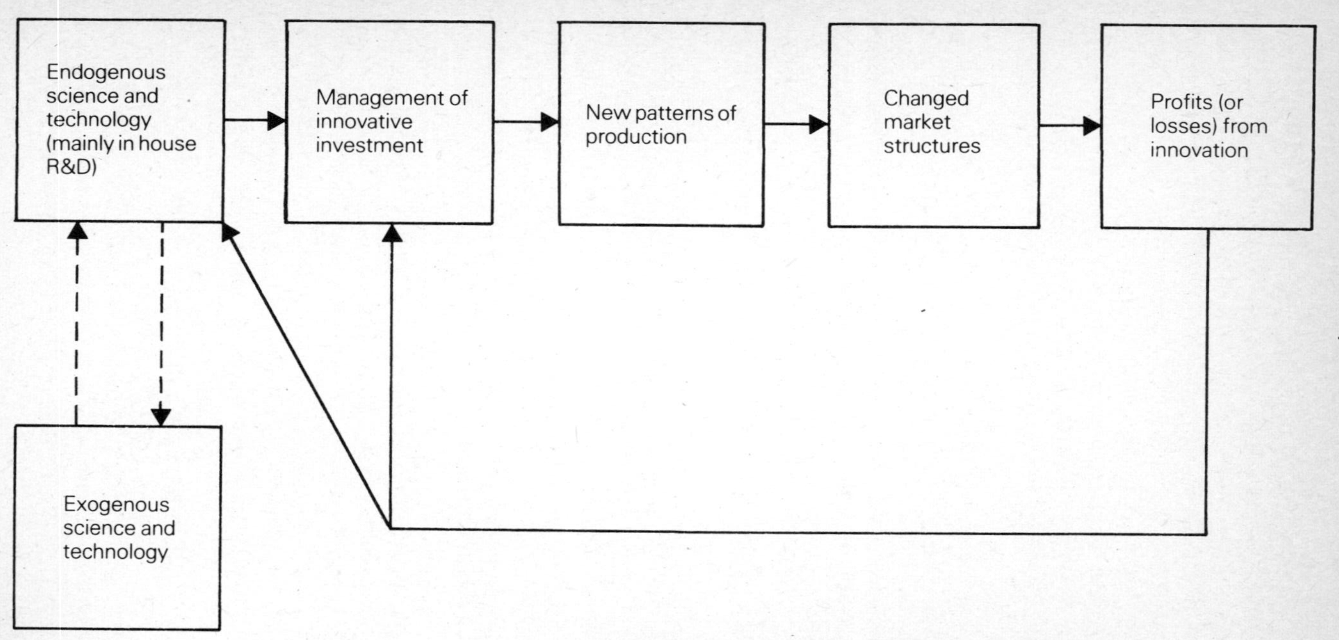

Figure 1.1 Schematic representation of Schumpeter's model of large-firm managed innovation (Mark II)

Source: Freeman, 1982, p. 40.

from the standardisation process and cost reduction efforts by innovators and swarmers, new producers would then be deterred from entering an already oligopolistic market where capital investments are huge and rates of return on these investments have been significantly diminished. Unless another *wave* of innovations is produced, this would eventually lead to the stagnation of the industry or its decline with increased international competition, whilst innovating firms direct their attention and efforts to more promising growth areas.

It is along these lines that in various parts of this book the development of the petrochemical industry and the roles of its major operators are studied.

1.4. The development stages of the petrochemical industry

The earliest plastic materials, cellulose, rayon and Bakelite, were invented and produced on a small commercial basis of individual inventor-entrepreneurs around the beginning of this century. Their production remained small and the quality poor due to the inability of the producers to improve the production processes and the performance of the products, whose markets and uses were limited.

Advances in the fundamental science of synthetic materials in the 1920s, particularly the contribution of Staudinger in Germany, who studied long-chain molecules, provided the theoretical basis of polymer science and laid down foundations for later developments and advances in products and production processes for plastics materials. Between the two World Wars, most of the major plastics materials were discovered or developed in the R & D laboratories of the largest chemical producers which were essentially involved in the production of synthetic dyestuffs and explosives. The German giant IG Farben was responsible for most of the innovation and development of synthetic materials (PVC, polyethylene, acrylics, nylon 6, PVCA, polystyrene, urea, etc.), followed by ICI (PVC, polyethylene, polyester) and Du Pont (PVC, polyester, nylon 66). IG Farben's leading position can be attributed to its strong R & D base and the subsequent concerted efforts prior to and during the Second World War period to attain German self-sufficiency in the production of synthetic rubber and other materials. These materials found ready but somehow limited applications during the War. In Germany, PVC and synthetic rubber were used mainly for waterproof wear and tyres respectively, while polyethylene and PVC in the United Kingdom were used to coat cables and wires for the radar systems.

In the post-war period, with the development of polymer science and the surge in R & D activities (mainly in the United States, United Kingdom and Germany), new products were discovered and the technical processes of production improved. Extensive research carried out by the leading chemical firms was responsible for most of the innovations during this period, but some of the important discoveries were made by independent researchers at universities, such as the Ziegler (Germany, 1950) and Natta (Italy, 1955) catalysts for the production of HDPE and polypropylene respectively. However, the development of products from bench-scale substances to pilot plant testing, followed by large-scale commercial production requires enormous R & D expenditure, industrial and technical know-how and large numbers of qualified personnel. These could be afforded only by the large chemical firms (such as IG Farben, ICI, Du Pont, Montecatini, etc.), which also had the experienced marketing channels to promote the sales of these products.

The expansion of the markets during the post-war construction and industrial boom period in the 1950s and 1960s, and the increased civil applications for synthetic materials, contributed to the rise in demand for these products, which found ready markets because the supply of conventional materials was limited and therefore more costly.

The *clustering* of the important innovations of new products and catalysts in the 1920s and 1940s, and in a second stage in the 1950s, allowed the large chemical firms to establish a leading position in the markets where they operated. A successful innovation leading to a new product gives the producing company a strong monopolistic position; it earns exceptional profits and can reinforce its position in the market before it faces competition by other producers. Nylon, for instance, accounted for a large proportion of the profits made by Du Pont for many years. ICI and IG Farben were in a similar position as regards polyethylene and PVC.

There is also evidence (Freeman, 1974) to show that the largest firms were the first to imitate or license a new product invented by another leading firm, as in the case of nylon 66 (Du Pont), nylon 6

(Hoechst), polyethylene PVC and acrylics (ICI, Du Pont, IG Farben). Hufbauer (1966) and Stobaugh (1968) demonstrated that the *imitation lag*, the time between the introduction of a new product into the market and its production by imitators, was shortest for producers with the highest innovation rate and a large market size. Also, when a major firm imitates an innovation, whether under licence or independently, it often attempts to introduce a major improvement into the product or to develop a better process for its production since this allows it to realise higher profits and to improve the prospects for its share of the market. This is evident from the present petrochemical industry where a variety of different processes exist for the production of most of the major plastics products as in HDPE (Solvay, Hoechst, Union Carbide, Phillips, Mitsubishi), LDPE (Union Carbide, ICI, ATO-Chimie, CdF-Chimie, etc.) and PVC (Solvay, ICI, Du Pont, Hoechst, B.F. Goodrich, Sumitomo, Wacker, etc.). Competition for market leadership has been the main incentive for the large chemical producers to develop their own technologies. Another aspect of this process is the speed of the technological progress. For example, only four years after the development of LLDPE by Union Carbide and DOW alternative processes were developed at Phillips, Mitsui, Amoco and CdF-Chimie.

All these developments point in the direction of the feed-back loop effect of Schumpeter's model (see Figure 1.1), showing the dynamic structure of the industry.

During the 1950s and 1960s, the industry was growing at extremely high rates and a swarming process had already been under way with production capacities spreading in most of the industrialised countries. The new growth cycle was characterised by a fresh wave of extensive R & D at universities and research centres of the large producing firms, which continued to lead in the number of innovations and patents taken out; basic innovations continued to be made (HDPE, PP, LLDPE) but most efforts were directed towards process innovations for the existing products – new production processes or scaling-up the capacities of plants. Rapid advances in polymer science over this period led to better products with wider market applications and as a result of the development of new plasticisers, copolymer additives and new catalysts the quality of products and the yield of production improved. Improvements were also made in the machinery for the fabrication and processing of plastics materials into end-products.

Despite the swarming process and the rapid increase in output, the number of producers remained small due to the large capital costs required for the new very large-scale plants and the oligopolistic structure of the industry. The swarming entrants were mainly the largest chemical and oil companies. The intensive R & D and technical advances, which continued to be dominated by the leading chemical companies, increasingly stimulated the demand for products which in turn led to another wave of technological and production improvements.

However, by the early 1970s, there were signs of decreasing returns to R & D expenditure; also, the indicators of patenting and scientific papers' publications were declining (Freeman, 1982, p. 98), as the technology was becoming mature and the process of standardisation increasing with the entry into the market of the engineering firms, which were able by now to supply their own technology to the would-be swarmers. The emphasis of R & D during this period shifted to the area of process development and control of chemical plants, energy saving and flexibility in the use of feedstocks. Significant advances were made in this field by the oil companies and the engineering firms.

The 1970s were also marked by the very large increase in the size of the petrochemical plants and, since the recent oil price rise (1979–80), most of the R & D research has been concerned with finding alternative production processes based on non-oil feedstocks such as methanol and SNG. On the other hand, the major chemical companies directed their attention to improving the performance of existing as well as new products to widen their market applications. The major chemical companies were concentrating their technological expertise on developing new branches of the industry to generate higher profits and sustain growth. Speciality products, pharmaceuticals and biotechnology are areas of future growth for the chemical industry, which are already under the focus of the research departments of the leading chemical producers.

During this period the markets for petrochemicals were becoming saturated and the industry's growth rates were slowing after penetration of most of the markets for conventional materials. Overcapacity has become widespread, particularly in Western Europe; profit margins have been squeezed as the number of producers increased and producing capacities multiplied while demand for their products started to increase at slow rates equivalent to the rates of increase in the use of the materials which they were replacing.

Another sign of the industry's increased maturity was the rise in input prices where the feedstock accounted for most of the production costs following the oil price rises; this also started a process of plant building and capacity expansion throughout the world and the industry was becoming increasingly internationalised. Stobaugh (1968, 1976) provides evidence that such a process has been taking place in the petrochemical industry, where for products arriving at the mature stage of their life cycle the production starts to shift slowly to those producers that have a comparative advantage, in terms of the main factor inputs, in the production of these products. The developments that are taking place in the oil-producing countries and the spread of construction of petrochemical plants throughout the world corroborate this view, as is argued in Chapter 4. Stobaugh also shows that prices of mature products were being forced down as the monopolistic producers were joined by imitating producers, and innovating producers were ready to license production processes only to firms over which they had considerable control. However, it is clear that despite the swarming process only a small number of producers operating in the major producing countries have an oligopolistic control over their markets. The recent developments in Western Europe reflect the oligopolistic practices carried out by the major petrochemical producers to reduce competition and improve prices (see Chapter 2 and Section 3.9 in Chapter 3).

1.5. The apparent factors affecting the demand for petrochemicals

A variety of technological, economic and social factors have led to the rapid expansion of demand for petrochemical products and the development of their industry. The development of large fluid catalytic crackers (FCC) for the refining industry in the United States during the early 1950s opened the way for the engineering firms to build similar plants for the petrochemical industry whose technology and operation are very similar to those of oil refineries. The increasing refining capacity throughout Europe and the United States in the post-war period made available large quantities of oil products which were readily used as petrochemical feedstocks.

The large profits made at the early stages of the introduction of new products into the markets were continuously reinvested to finance more modern and large plants. This enabled petrochemical products to be produced in large quantities at competitive prices in close proximity to the expanding markets. The interdependent and dynamic structure of petrochemical production, where different products can be produced in a large 'complex', encouraged the development of market uses for the by-products and at times allowed cross subsidy of production (for example, ethylene and by-products from naphtha).

The development of a processing industry for plastics materials (moulding, injection, extrusion, etc.) and synthetic fibres and the improved properties of these materials, where they were produced in various grades and with different specifications, allowed the production of many kinds of products in different shapes and forms (for example, sheets, films, drums, pipes, tyres, fibres, etc.) for numerous applications which greatly expanded their end-use markets and made them more acceptable to the consumers. On the other hand, industrial producers were willing to increase their use of synthetic materials as these ensured a consistent supply of cost-effective materials produced at home in large quantities reducing dependence on imported conventional materials, whose prices and supplies fluctuated frequently. Also, developments of polymer science and process technology led to new products tailor-made for their end uses, which speeded up the penetration of new markets.

Finally, expanding industrial production, particularly in food processing and packaging, car production and the building industry, together with the rise in the standard of living, contributed to the rapid rise in demand for petrochemical products in the industrially developed countries over the last two decades.

In spite of the oil price rises of the 1970s, which caused sharp increases in the costs of feedstocks and raw materials for petrochemical products, especially plastics, the demand for these materials continued to go up, albeit at lower rates than before since the higher oil prices were translated into higher production costs of products for which plastics were substituted.

Plastics materials will continue to substitute conventional materials such as steel, copper, glass, paper, etc., in various applications and their potential for increased penetration, throughout the world, into existing and new markets is still enormous. A potential future market for plastics is the aviation industry since, in addition to the lightness, mouldability and the strength as great as steel, which these plastic composite materials offer, they are not detectable on radar screens, which makes them of particular interest to military producers.

The future implications of higher energy costs and prospects for increased plastics consumption in two already large markets, the car and the construction industry, are briefly discussed below. But first the supply and demand situation for the major petrochemical products, which existed over the last two decades, and the projected trend until 1990 is dealt with.

1.6. The supply and demand pattern of petrochemical products

1.6.1. Overview

The production and consumption of petrochemicals is concentrated in three regions of the world, North America, Western Europe and Japan, which together account for about 80 per cent of world production and more than 75 per cent of consumption. The concentration is highest for plastics materials, which decreases for man-made fibres and synthetic rubber due to the increasing participation by developing countries in the production in the former and East Europe and the Soviet Union in the latter.

The numerous petrochemical products can be classified into three types of products: basic or bulk, intermediate and end or final petrochemicals, according to their function and end use (shown in Figure 1.2).

Only the major products of these materials are studied in this book, particularly the large tonnage thermoplastics (PE, PP, PVC, PS), whose production has grown from a few thousand tonnes in the 1950s to a few million tonnes each throughout the world at present. These materials have a wide variety of applications. The percentage consumption for downstream and final use of some of the products is summarised in Table A1 in Appendix A.

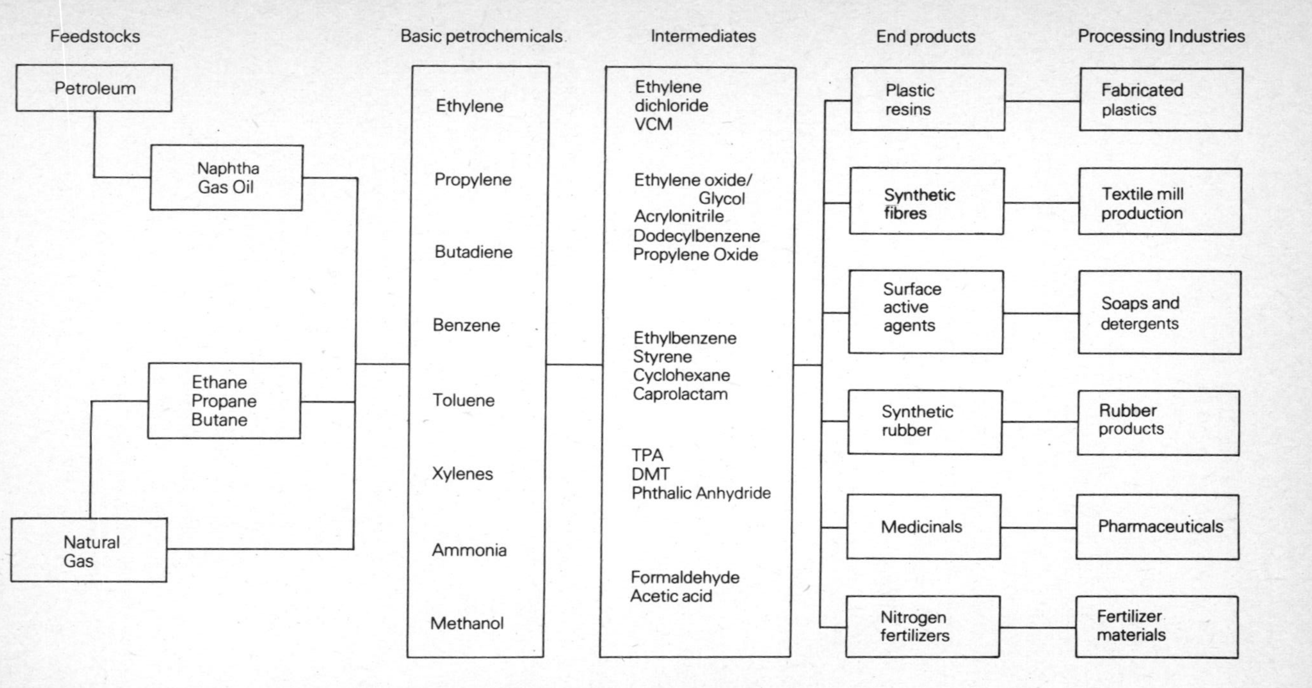

* Plastic resins include HD polyethylene, LD polyethylene, polypropylene, polystyrene, PVC ABS vinyl acetate
† Synthetic fibres include acrylic fibres, nylon (polyamide) fibres and polyester fibres
‡ Synthetic rubbers include polybutadiene, SBR, polyisoprene, butyl rubber
§ Nitrogen fertilizers include urea, ammonia nitrate etc.

Source: UNIDO, ID/WG. 336/2, March 1981, p. 7.

Figure 1.2 Different stages of production in the petrochemical industry

1.6.2. Basic petrochemical products

The most important basic petrochemical products are ethylene, propylene, butadiene and benzene. Ethylene – the back-bone of the petrochemical industry – is the largest single product produced. Its world production reached 40 million tonnes in 1979, with 18 million tonnes produced in North America while West Europe and Japan accounted for 15 million and 6 million tonnes respectively.

Tables A2 to A5 (Appendix A) show the trends of production of ethylene, propylene, butadiene and benzene. The corresponding growth rate figures point to similar trends of production development for all the products in the three largest producing regions, where throughout the 1960s and until 1974 very high rates of growth were achieved. These eased slightly after 1974, rising at lower rates than in the early stages of the products' development, where production was increasing from a much lower base.

The average annual growth rates differed in absolute value for each product and from one producing region to another. For the two larger volume products, ethylene and propylene, they were higher than those of butadiene and benzene. Production was increasing more rapidly in Japan and West Europe than in North America, as illustrated by the average annual growth rates, which for ethylene and propylene were about 30 per cent in Japan, 22 per cent in West Europe and 12 per cent in North America respectively; while those for butadiene and benzene were slightly lower but followed the same order.

After 1974, however, the rapid growth rates slowed down for all four products and production was increasing at equivalent rates in the three regions, which averaged between 5–8 per cent per annum. It is also significant to note that by 1979 the production levels of those products in Japan were equivalent to about half of the total West European output, which in turn was approaching that of North America, as can be clearly seen from the tables.

The production capacities for ethylene, propylene, butadiene and benzene, shown in Tables A6 to A9 (see Appendix A), are indicative of the supply pattern of basic petrochemical products. The tables show that production capacities were increasing in step with the rising demand. The average utilisation rates for these capacities lay between 80 per cent and more than 90 per cent during the rapid growth period until 1973, and declined during the rest of the 1970s to about 70–75 per cent.

The balance that existed in the 1960s between supply and demand, mainly as a result of very high growth rates in demand and a small number of producers, had ceased to exist since 1976, when demand growth rates declined from the previously high rates of the 1960s and capacities were increased rapidly over a short period, following the oil crisis of 1973, with the entry of oil and state-owned companies into the petrochemical markets in a substantial way.

Many plants were built in 1976 and many more continued to come on stream throughout the rest of the 1970s, although the recovery in demand over the 1976–9 period was short-lived and the expectations of high rates of growth in demand did not materialise.[1] The long duration of the recession has depressed the demand for petrochemicals which, together with the increased number of producers (although still small) and the very large capacity of the newly built plants, off-set this balance, particularly in West Europe, where old plants that were totally amortised for a long time were still in operation, adding to the overcapacity problem.

The low growth rate figures registered in the early 1980s can be attributed to the recession in general and the sluggish growth of industrial output particularly for the industries which are the major users of these products. But since the early 1970s there were signs that the industry's rates of growth were slowing down as most markets were already deeply penetrated, particularly in Japan and Western Europe.

But what about the future trends in the growth of basic petrochemical products in the major producing regions? Projecting future demands, in general, is a very hazardous business unless based on a detailed analysis of the industrial structure and an accurate estimation of the economic performance of the market under study, which is discussed in more detail in Chapter 3.

However, it is widely accepted now that the growth of petrochemical output will slow down in the future in the United States, Japan and Western Europe, where the industry is showing increasing signs of maturity. Growth of production of petrochemicals is expected to be more in line with the level of industrial growth. OECD estimates predict that the growth of production of basic petrochemicals in the 1980s will be equivalent to the growth of GDP in Japan and Western Europe, while in the United States it would be 1.5 to 2 times the rate of growth of GDP.[2]

Estimates made by UNIDO (ID/WG 336/3, 1981) did not differ significantly from those made by OECD, CEFIC and others but

Table 1.1 Demand for basic petrochemicals in industrialised countries in 1975 and 1990 (million tonnes)

	ethylene		*propylene*		*Butadiene*		*Benzene*		*xylene*	
Region	*1975*	*1990*	*1975*	*1990*	*1975*	*1990*	*1975*	*1990*	*1975*	*1990*
N. America	9.80	18.65	4.40	9.32	1.50	2.27	3.74	6.80	1.32	2.38
W. Europe	7.90	12.40	4.10	7.07	0.81	1.36	3.26	4.85	1.09	1.65
Japan	3.40	3.94	2.30	3.27	0.47	0.84	1.48	2.00	0.65	0.88
USSR and E. Europe	2.00	4.60	1.20	3.05	0.45	0.83	2.15	3.61	0.60	0.88
Other countries	0.25	0.58	0.12	0.27	0.10	0.09	0.08	0.17	0.02	0.02
Total	23.25	40.17	12.12	22.98	3.33	5.39	10.71	17.43	3.68	5.82

Source: UNIDO, ID/WG. 336/3, May 1981, UNIDO IS.427, Dec. 1983; SRI.

were rather optimistic. However with the weakening in world economic activity the same sources later adjusted downwards their previous forecasts.

A subsequent UNIDO study (UNIDO, I.S. 427, 1983) [3] together with another source, SRI, whose projections seem to be more reasonable, have formed the basis of the estimates presented here. However, some industrial sources (BP and others) have taken a more pessimistic view about the future development of the petrochemical industry. Their estimates are unrealistically low; but their increased activities in the petrochemical markets do not seem to reflect their cautious forecasts.

It should be stressed, however, that all these projections (including OECD and CEFIC), rely heavily on the general economic indicator of Gross Domestic Product (or its major components), while the course of industrial development in the long term may affect the growth of demand for petrochemicals disproportionately higher, or lower, than the level of GDP growth, as we shall argue in a later section of this book. Other factors, such as international trade, exchange rates, market saturation and developments in the petrochemical industry in other regions of the world, would inevitably affect the production growth of petrochemicals in the major producing regions.

Table 1.1 reveals that the growth of demand for the four basic

Table 1.2 Planned increase in production capacity for selected petrochemicals in industrialised countries, 1979–1990. (million tonnes)

	Estimated increase in demand 1979–1990	*Estimated increase in capacity 1979–1984*	*Excess capacity available in 1984**	*Additional capacity required in 1984–1990**
Basics				
Ethylene	5.30	2.94	3.87	2.95
Propylene	4.47	1.64	1.43	3.16
Butadiene	0.79	0.13	0.30	0.70
Benzene	1.41	3.04	2.98	—
Xylenes	0.38	0.47	0.37	—
Methanol	4.25	2.45	—	2.29
Plastics				
LDPE**	2.58	−1.05	−1.24	3.42
HDPE	2.04	3.76	1.90	—
PP	2.56	0.91	−0.23	1.83
PVC	2.21	1.85	—	0.73
PS	0.50	0.60	0.40	—

* Assuming plants operate at 80 per cent capacity
** Includes LLDPE.

Sources: Tables 1.5, 1.6, 1.1

petrochemicals is estimated to range between 2 and 4 per cent per annum until 1990. The year-to-year figures, however, may be higher or lower than this average and they would vary from one country to another during this period.

The development of new projects in the Scandinavian countries, Greece, Spain and Portugal will stimulate demand to higher rates than those in the other Western European countries and will account for a significant proportion of the additional capacities required to meet the Western European demand.

In the United States and Canada, the demand for petrochemicals is expected to be higher than in Western Europe and Japan since there are still potentially large markets to be penetrated. Also industrial output there is expected to rise more rapidly than in Western Europe; whereas Japan seems to be emphasising more promising industries, such as biotechnology and electronics, and is slowly moving away from energy-intensive industries.

Although the growth of demand for petrochemicals until 1990 is expected to be modest in the industrialised countries, considerable additional capacities are required to meet this demand which for ethylene alone amounts to about 40 million tonnes in 1990 (see Table 1.1). The additional capacities would be spread over two five-year periods as shown in Table 1.2. There are already plans for a total ethylene capacity of about 5 million tonnes that will come on stream by the mid 1980s in Canada (1 million), Alaska (1 million), Scandinavia (1 million), United Kingdom (1 million) and the remaining European countries.[4]

1.6.3. Major final petrochemical products

Plastics materials account for more than 50 per cent of the total end petrochemical products. Their production is dominated by thermoplastic products, namely PE, PVC, PP and PS,[5] and their supply and demand development pattern is similar to those of the basic petrochemical products, for which in the 1960s and up until 1973 world demand grew considerably and was mainly concentrated in the three main producing centres, United States, Western Europe and Japan. By the end of this period, however, a certain decline in the demand growth rates for these products was being felt. The 1974–5 period was characterised by a drop in demand throughout the world, which recovered in 1976 to a level close to that of 1973–4 but was now growing at slower rates than before, reflecting an increasing saturation of the markets, especially in Western Europe and Japan.

The trends in the growth of production of the four major plastics materials PE, PVC, PS and AP are depicted in Figures 1.3, 1.4, 1.5 and 1.6 respectively, and that of synthetic rubber in Figure 1.7. These show that the production growth pattern of all the products is similar, with production increasing rapidly until 1973, but thereafter becoming irregular and slower than before as a result of the economic stagnation throughout the world, especially in the industrialised countries.

The rates of growth varied slightly, depending on the type of product and the demand structure in each country. However, in the industrialised countries in general, the rates of growth for the major plastics after 1974 averaged between 5 and 11 per cent per annum, but were higher for recently developed products such as PP and HDPE which were still growing at rates of over 12 per cent. Also, production in North America, particularly the United States, continued to grow at higher levels than in Western Europe; and in

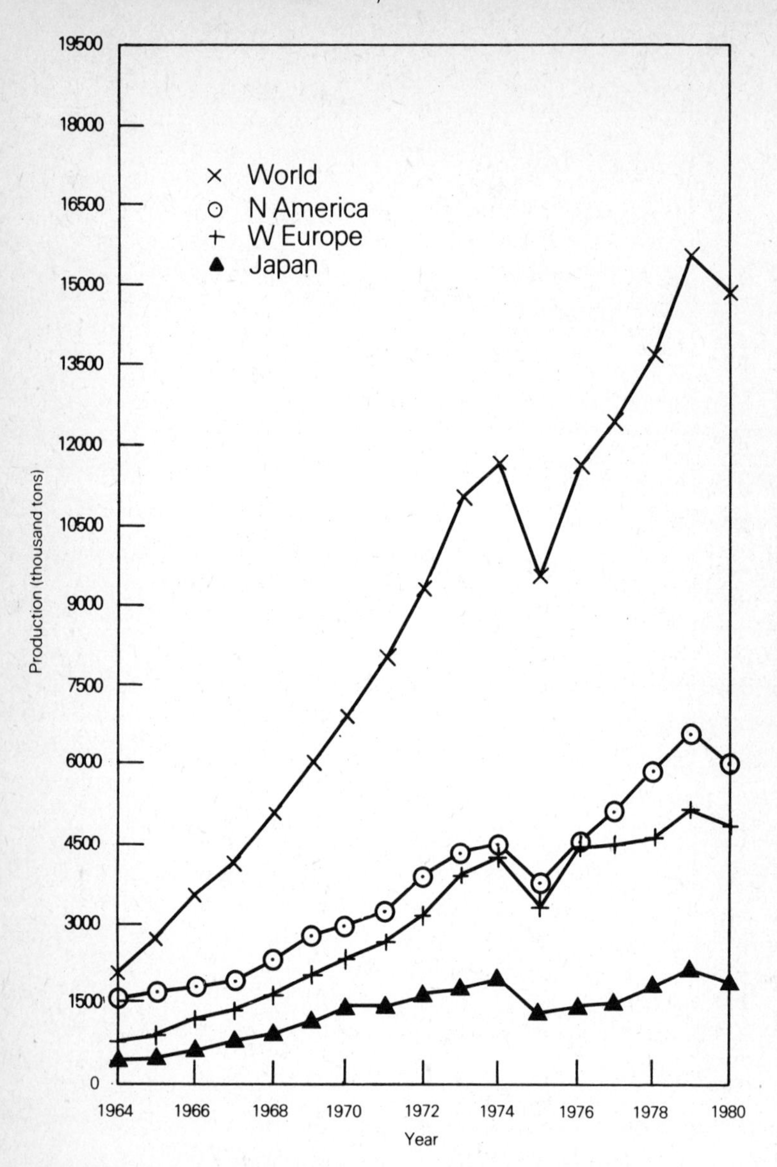

Figure 1.3 Polyethylene production by the main producing regions and the total world

Source: Based on data obtained from *United Nations: Yearbook of Industrial Statistics,* various issues.

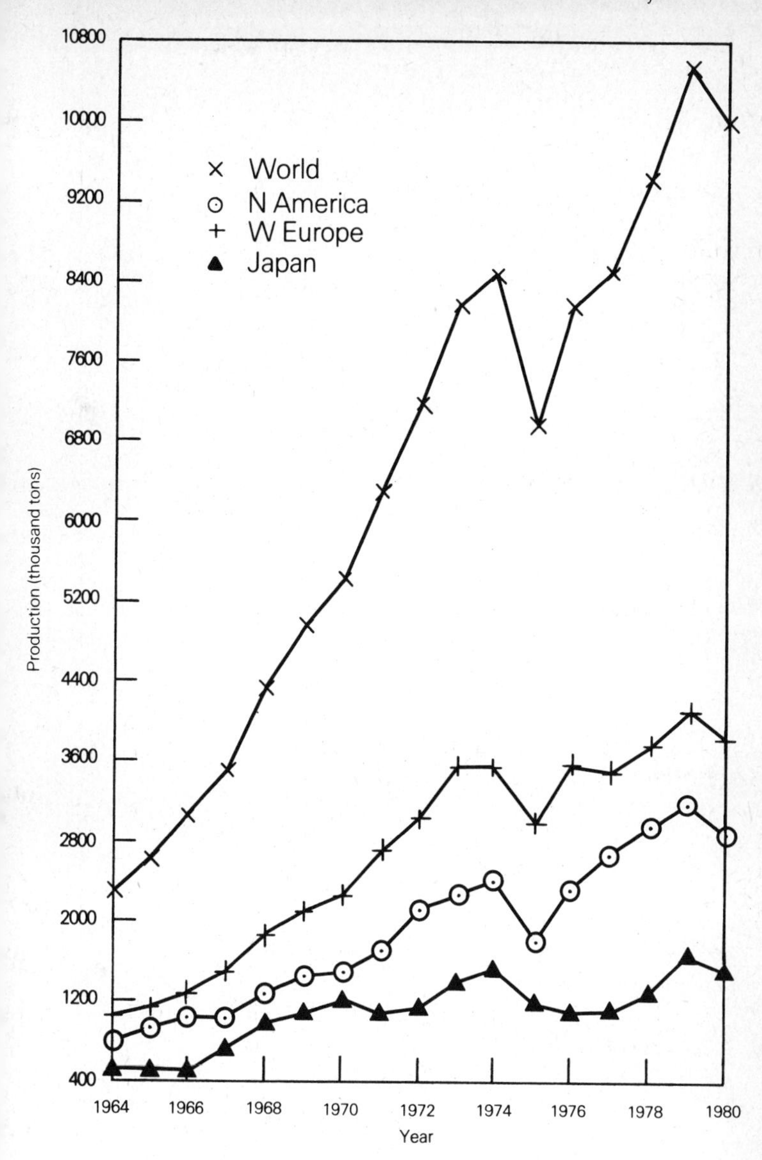

Figure 1.4 Polyvinyl chloride production by the main producing regions and the total world

Source: As Figure 1.3.

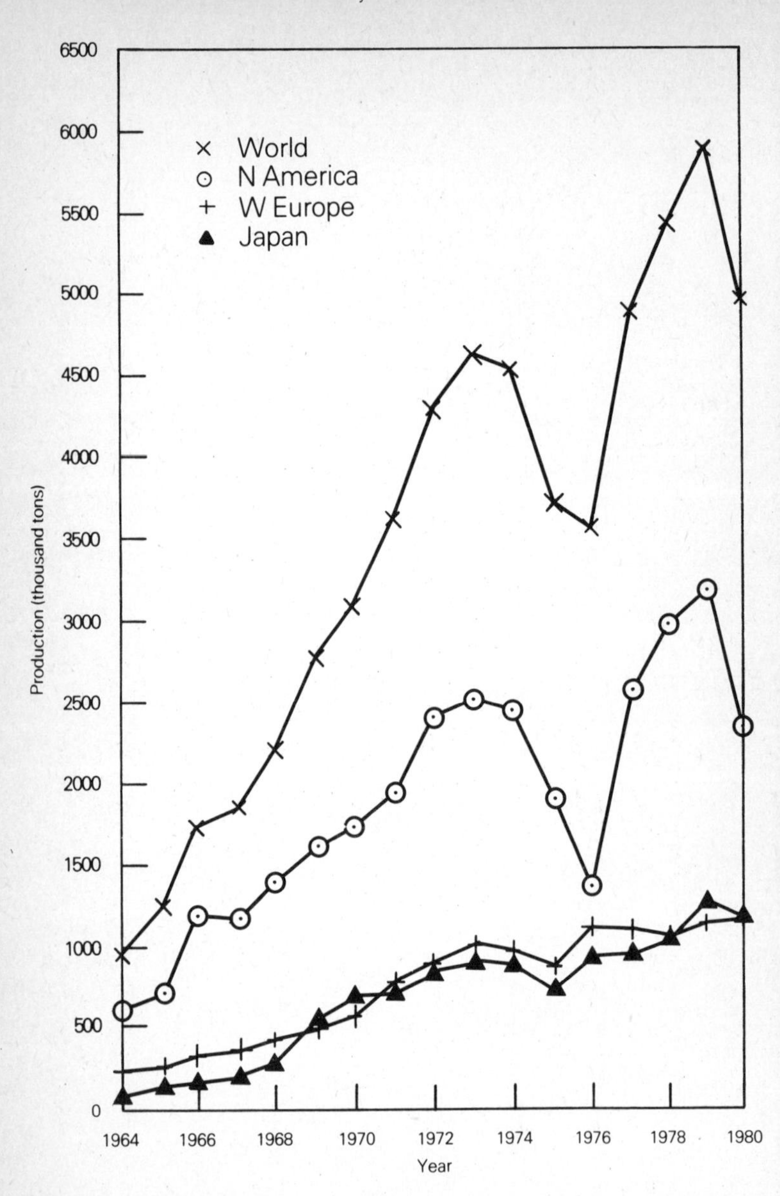

Figure 1.5 Polystyrene production by the main producing regions and the total world

Source: As Figure 1.3.

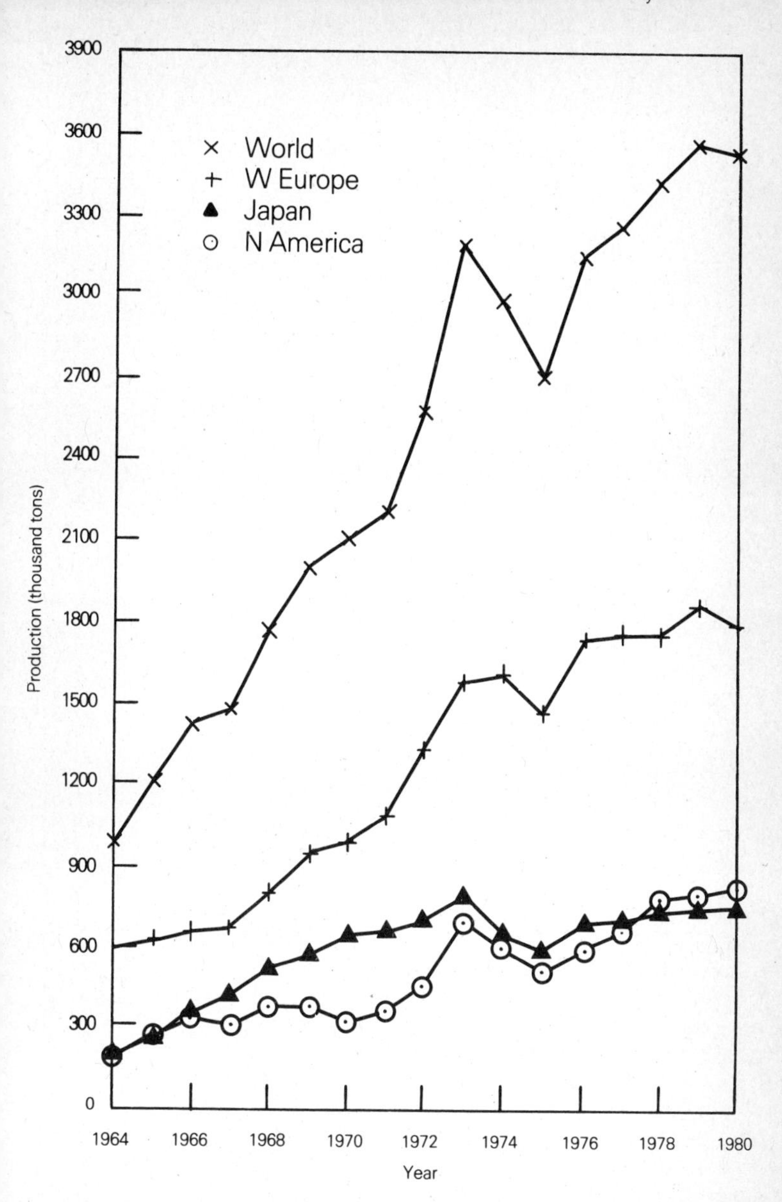

Figure 1.6 Amino plastics production by the main producing regions and the total world

Source: As Figure 1.3.

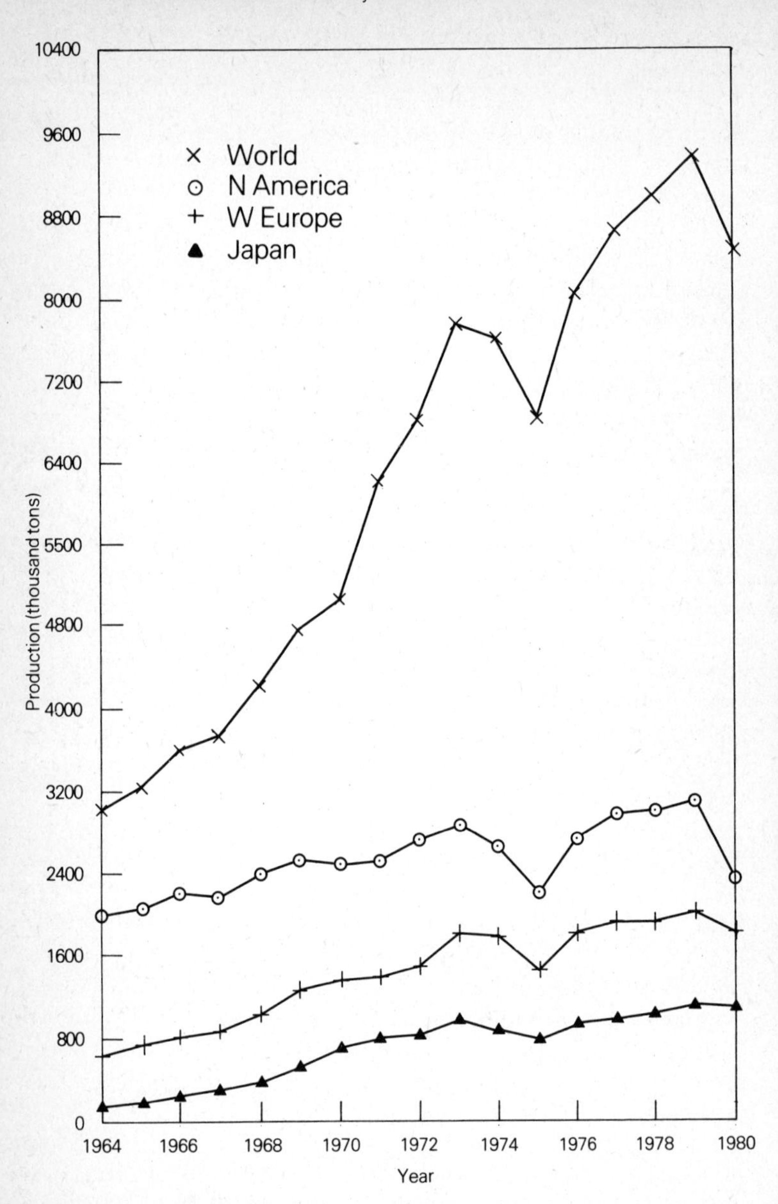

Figure 1.7 Synthetic rubber production by the main producing regions and the total world

Source: As Figure 1.3.

West Germany the situation was better than in other European countries.

In 1980 there was another drop in production in all regions which continued into 1982, as industrial production declined or failed to pick up sufficiently in most of the industrialised countries.

The 'world' production curves in the post-1973 period show a higher rate of growth than those of the main producing regions for all products, reflecting the expansion of production capacities and increasing demand throughout the world, particularly in Eastern Europe and the newly industrialised countries, thus depicting a pattern of growth consistent with the product life cycle concept for maturing industries.

The production of synthetic rubber has been less concentrated in the main producing regions (North America, Japan and Western Europe) than other products, where production capacities existed for a long time in various countries, particularly in the Soviet Union, Eastern Europe, Brazil and India, and which continued to rise rapidly after 1975 at rates much higher than those of the three major producing regions. This contributed to the very high growth rates achieved by world total production (Fig. 1.7). Production declined after 1979, the largest drop in output was registered in the United States, which continued into 1982 as a result of the drop in car production, which accounts for most of the use of synthetic rubber in that country.

With the exception of AP, the output of all major plastics, particularly PS, PP and PE, have since 1975 been growing at higher rates in North America than in Western Europe and Japan. However, the level of production of PVC and AP is still higher in Western Europe than in North America (see Figures 1.3–1.6). The slight variation in the pattern of production growth of plastics materials between the major producing regions can be explained by the changing structure of demand for the individual plastics products in each country, resulting from the pattern of their end use in each market.

The changing structure of production in the three major producing regions is displayed in Table 1.3. A striking feature of the structure of the Japanese market in 1980 was the more even distribution of the market shares among the individual products than in Western Europe and North America. In the three regions, however, the shares of HDPE and PP have been increasing steadily

Table 1.3 The structure of production of major plastics in the major producing regions

	Thousand tons per year			*Percentages*		
Region	*1971*	*1975*	*1980*	*1971*	*1975*	*1980*
W. Europe						
LDPE	2,634	2,370	3,467	40	31	32.6
HDPE		862	1,283		11	12
PP	550	730	1,077	8	9.5	10
PVC	2,634	2,869	3,673	40	37.5	34.5
PS	760	838	1,132	12	11	10.6
Total	6,578	7,669	10,632	100	100	100
N. America						
LDPE	3,109	2,483	3,642	42.8	30.5	29.3
HDPE		1,164	2,269		14.3	18.2
PP	607	863	1,516	8.3	10.6	12.2
PVC	1,658	1,764	2,771	22.8	21.7	22.3
PS	1,889	1,855	2,243	26.0	22.8	18.0
Total	7,263	8,129	12,441	100	100	100
Japan						
LDPE	1,340	907	1,018	36.2	24.5	19.0
HDPE		383	861		10.3	16.0
PP	627	594	927	17.0	16.0	17.3
PVC	1,034	1,125	1,429	28.0	30.4	26.6
PS	695	690	1,129	18.8	18.6	21.0
Total	3,696	3,699	5,364	100	100	100
World						
LDPE	7,773	9,235	14,479	40.3	42.0	44.0
HDPE						
PP	1,879	2,374	3,933	9.7	10.8	12.0
PVC	6,094	6,760	9,668	31.6	30.7	29.4
PS	3,514	3,597	4,802	18.2	16.4	14.6
Total	19,260	21,966	32,882	100	100	100

Source: *UN Yearbook of Industrial Statistics*, various years.

at the expense of LDPE and PVC (Table 1.3). Production of LDPE has been increasing to approach that of PVC in Western Europe; while in North America, LDPE has been the largest single product in production since the early 1970s. In Japan, PVC still plays a dominant role as the leading plastics product, but PS production

Table 1.4 Production of plastics materials in the three major producing regions as a percentage of world total output

	1971				*1980*			
	PE	*PP*	*PVC*	*PS*	*PE*	*PP*	*PVC*	*PS*
W. Europe	34	29	43	21.5	33	27.0	38	23.5
N. America	40	32	27	54	41	38.5	29	47.0
Japan	17	33	17	20	13	23.5	15	23.5
Total (Japan + N. America + W. Europe)	91	94	81	95.5	87	89	82	94

Note: Based on Table 1.3.

has been increasing steadily to capture an equivalent share of the market as LDPE.

Table 1.4 compares the changing pattern of production of each product and the corresponding share of each producing region. The drop in the three regions' percentage share of the world total production for most products (see last row of the table) indicates the increased participation in petrochemical production by other countries. The demand for the major plastics products is expected to continue to grow at modest rates in the industrialised countries until 1990.

As in the case of basic petrochemicals, it is difficult to project reliable demand figures for individual end products without detailed analysis of the industrial market structure and the prospects of economic development in the context of an econometric model which takes into consideration other aspects of the market under study related to the petrochemical industry and the major outlets for its products. This point is covered more fully in Chapter 3. However, estimates based on the UNIDO model for petrochemicals and SRI are presented in Table 1.5, which reveals that the estimated average annual growth rates of demand range from 3–5 per cent for the major plastics products over the period 1985–90, while that of PP is expected to be slightly higher at about 6.5 per cent. These figures may vary from country to country,

Table 1.5 Growth of demand for selected petrochemical products (per cent per annum)

Petrochemical product	*World total*			*Industrialised countries**			*Developing countries*		
	1975–1979	*1980–1985*	*1985–1990*	*1975–1979*	*1980–1985*	*1985–1990*	*1975–1979*	*1980–1985*	*1985–1990*
Basic petrochemicals									
Ethylene	11.2	1.5	3.7	10.4	1.0	2.8	24.0	7.5	7.4
Propylene	11.8	3.0	3.7	8.7	2.7	3.4	26.0	7.0	6.3
Butadiene	8.8	2.9	5.1	7.8	2.4	4.7	18.9	7.5	7.8
Benzene	11.0	1.8	2.3	10.8	1.4	2.0	12.8	7.2	5.9
Xylene	11.8	2.7	2.6	9.3	2.0	1.7	41.4	9.5	8.5
Methanol	10.7	3.1	4.4	10.1	2.8	3.8	22.9	7.4	10.0
Thermoplastics									
LDPE†	10.3	4.3	5.0	11.3	3.5	4.8	19.1	8.4	5.7
HDPE	17.9	5.9	5.1	16.9	5.5	5.0	24.4	8.3	5.4
PP	20.4	6.4	6.6	19.0	6.0	6.5	28.7	8.2	7.4
PVC	11.4	4.0	3.9	10.6	3.3	3.4	16.1	7.8	6.3
PS	11.9	3.6	3.5	11.2	3.1	3.2	18.1	6.7	5.2

Sources: UNIDO, ID/WG. 336/3, May 1981, p. 54; UNIDO, IS. 427, Dec. 1983; SRI

* Including Eastern Europe and the USSR.

† Includes LLDPE

Table 1.6 Estimates of capacity to produce selected petrochemicals in industrialised countries (million tons)

Petrochemical products	*Japan*		*W. Europe*		*N. America*		*Soviet Union and E. Europe*		*Other countries*		*Total*	
	1979	*1984*	*1979*	*1984*	*1979*	*1984*	*1979*	*1984*	*1979*	*1984*	*1979*	*1984*
Basic petrochemicals												
Ethylene	5.98	5.43	16.11	14.58	18.48	20.57	3.66	6.32	0.45	0.64	44.60	47.54
Propylene	3.73	3.76	8.99	8.22	9.96	11.38	1.77	2.77	0.26	0.22	24.71	26.35
Butadiene	0.74	0.79	2.19	2.08	2.31	1.88	0.44	1.00	0.06	0.12	5.74	5.87
Benzene	2.80	2.89	6.71	7.15	8.62	9.53	2.26	3.70	0.14	0.30	20.53	23.57
Xylenes	1.11	1.86	1.98	2.35	3.05	6.30	0.90	1.70	0.09	0.09	6.87	7.34
Methanol	1.30	0.53	3.08	2.80	5.51	7.09	2.25	3.80	0.12	0.50	12.27	14.72
Thermoplastics												
LDPE*	1.50	1.40	6.29	5.53	4.17	3.81	1.46	1.61	0.20	0.22	13.62	12.57
HDPE	0.90	1.07	2.43	2.86	2.82	5.28	0.30	0.88	0.15	0.27	6.60	10.36
PP	1.15	1.39	2.20	2.40	2.54	2.55	0.35	0.81	0.18	0.18	6.42	7.33
PVC	1.86	1.84	5.39	5.47	3.96	4.48	1.33	2.60	0.31	0.31	12.85	14.70
PS	0.92	1.04	2.89	2.82	2.73	3.01	0.36	0.61	0.07	0.09	6.97	7.57

* Includes also LLDPE capacities.

Sources: UNIDO, ID/WG. 336/3, May 1981, p. 64; SRI.

depending on the market structure, the products' present level of saturation and the level of industrial output.

Furthermore, it will be difficult to determine precisely the level of production of individual products as the facility to substitute between PP, PVC and HDPE is increasing in many applications, since the economy does not require particular products but the economic application they fulfil.[6] The future demand for these products will increasingly depend on their performance, price and availability.

The existing production capacities in the major producing regions have been capable of meeting the demands until 1984. Substantial additional capacities will be required beyond 1985, particularly for LDPE (in fact for the new type LLDPE) and PP (see Tables 1.6 and 1.2) in the industrialised countries. The large bulk of these products will continue to be supplied within each producing region until 1990. Some imports into these markets are, however, expected to originate from plants that have come onstream in 1985 in oil-producing countries such as Mexico and the Middle East countries. So the situation could change beyond 1990 with increased international trade in plastics materials, and oil-rich petrochemical producers could play an important role. The case for these developments is argued in Chapter 4.

1.7. Plastics materials and the energy problem

Over the period from the 1950s until the early 1970s the use of plastics materials increased at average annual rates of about 15 per cent, mainly as they were increasingly substituting conventional materials such as steel, glass, paper, wood, etc., in various applications. Prices were declining in real terms until 1973 as a result of cheap raw materials and the technological advances made in production processes. However, the sharp oil price increase of 1973 caused a significant rise in the price of plastics products as they are made almost totally from oil products.

The oil price rise and its impact on the cost structure of plastics materials (discussed in Chapter 2) raised doubts about the future of plastics products and it was suspected that their main users would shift back to the use of conventional materials. In fact, all non-plastic products began to rise in price during 1975–7 as the effects of the oil price rise worked their way through the economies and were felt in all industrial products and processes. Paper, tinplate and a few other materials, however, reacted to the new situation at the same time as plastics.

Between 1973 and 1979, prices for most plastics (wholesale price indices for the United Kingdom) increased from 140 per cent to 190 per cent.[7] The prices of paper, glass, steel and zinc also increased. However, the highest price increase, of over 200 per cent, was registered by aluminium. In terms of actual money prices, LDPE was still cheaper than paper, and in many other cases it was and will continue to be the *value* of plastics – their handling and performance – that is significant. For instance, a refuse collection worker would carry a total weight of 16,500 kg/day when handling metal dust bins compared with only 9,200 kg/day when polyethylene bags are used.[8]

Plastics materials have replaced conventional materials and are expected to continue doing so in the future. In terms of energy use, plastics are in most cases more efficient than their alternatives. Table 1.7 shows the energy requirements to produce the basic materials for unfinished products, in which the energy content of plastics materials in terms of tonnes of oil equivalent (TOE) is significantly lower than those of aluminium, steel, tinplate and copper on a volume basis due to the lower density of plastics materials; since in the end plastics products would replace the function played by their substitutes, which in most cases is more related to the size and shape of the replaced item than to its weight.

Comparisons of the energy content of fabricated products (Barry *et al.*, 1977) for final use reflect more accurately, than the (TOE/tonne of material), the advantage of plastics products in the use of energy over most of their conventional competitors in the production of pipes, bottles, films and bags. However, two points about these calculations should be clarified before proceeding any further in assessing the prospects of plastics materials. First, the energy content for the different materials based on TOE values cannot always be compared directly in cost or money terms as they relate to different forms of energy sources (coal, gas, oil, naphtha, etc.) which have different market prices. Second, there are other production costs which are not taken into account in these calculations, such as labour costs, investment costs, the extent of the economies of scale involved and the successive number of processing stages required before the final product is obtained.

Table 1.7 United Kingdom energy requirements in the production of basic materials

		Oil equivalents by weight (TOE/tonne)			*Total TOE per tonne equivalent*
Material	*Density g/cm³*	*TOE per tonne for feedstock*	*TOE per tonne for conversion*	*Total TOE per tonne of basic material*	*kcal per cm³*
Aluminium	2.7	—	5.6	5.6	158
Steel billet	7.8	—	1.0	1.0	82
Tinplate	7.8	—	1.25	1.25	102
Copper billet	8.9	—	1.2	1.2	112
Glass bottles	2.4	—	0.6	0.6	11
Paper and board	0.8	—	1.1	1.1	9.3
Polystyrene	1.07	1.3	1.88	3.18	36
PVC	1.38	0.55	1.4	1.95	28
LDPE	0.92	1.11	1.13	2.24	22
HDPE	0.96	1.13	1.2	2.33	24
PP	0.90	1.17	1.38	2.55	24

Source: C. Vowles (1979).

Nevertheless, plastics materials still enjoy a cost advantage over their substitutes in most applications and, if anything, the oil price increases have improved the prospects for the long-term growth of most plastics products.[9] Also, price for various sources of energy may become more related to their calorific value; for example some natural gas producers (Sonatrach of Algeria) link contract gas prices to those of oil. This process would be further strengthened as conventional energy sources become more scarce in the future.

Industrial circles, academic energy specialists and the IEA[10] tend to agree that oil and energy prices in general will continue to rise in real terms in the long run, by between 3–8 per cent annually until the year 2000 (Motamen, 1983). British Petroleum[11] and the IEA predict that by the end of this decade demand for energy, particularly oil, would outstrip supply, which in the long run could have an effect in reducing the price differentials that exist now

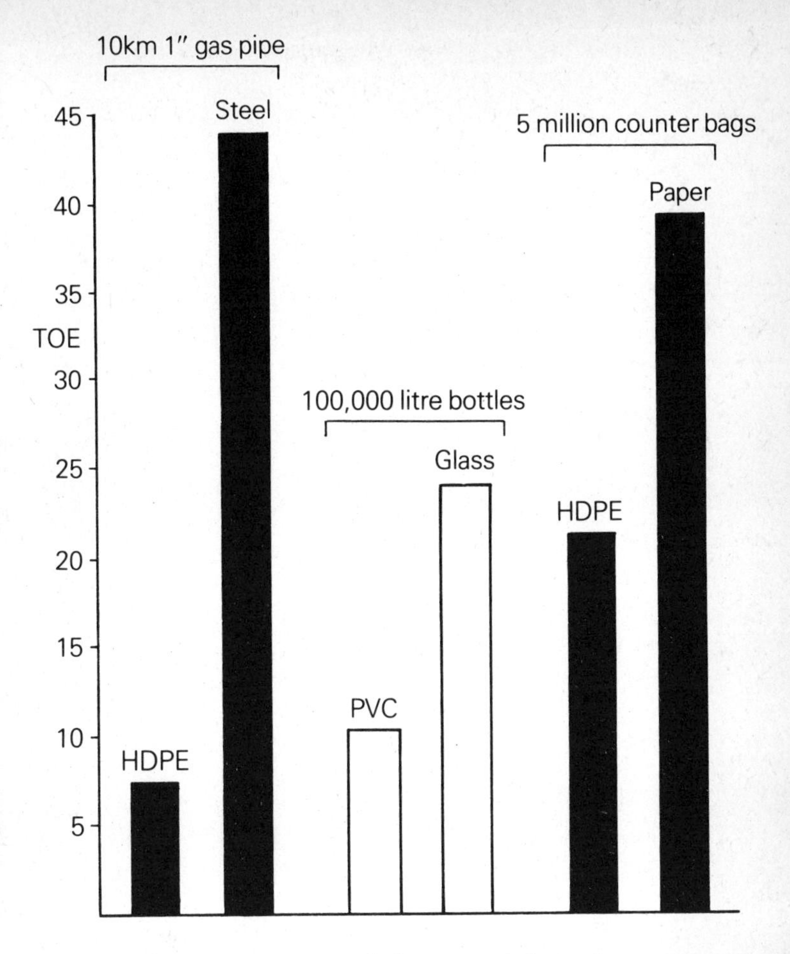

Figure 1.8 Energy content of plastics and their alternatives

Source: D. Sharp and T.F. West (eds), *The Chemical Industry*, 1982, p. 189.

between the different energy sources (coal, gas, oil products, electricity) when the technology to exploit their most useful potential is developed. Prices would become more related to the calorific value of the energy source, its cost of production (or that of its substitutes) in the final form in which it is used, and its availability.

This situation would further strengthen the position of plastics materials as they have a lower *energy content* in their final form than most of their substitutes. Plastics materials are more energy efficient not only in manufacture but also in transportation over long distances because they are light (e.g. bottles and pipes).

The production of plastics products is usually less labour intensive than that of conventional materials. Also, the application of the end-use plastics products is in many cases less labour intensive which makes them in effect more attractive to manufacturers and employers, needing to reduce overall costs of production.

In addition to their light weight, plastics offer a combination of physical strength, electrical and thermal insulation, resistance to corrosion and ease of fabrication into different shapes and sizes. So it is not only the price of plastics that gives them an advantage over conventional materials, but also their superior qualities which make them of better *value* in the applications they fulfil.

1.8. The future prospects for plastics in the car and building industries

The car and the building industries already account for a large share of plastics output but the intensiveness of plastics use, that is, the amount of plastics used as a percentage of total materials consumption, by the two industries is still very small. However, there remains a very large potential for increased plastics penetration in these two sectors.

Previously, only small amounts of plastics were used in cars which, excluding paint and tyres, amounted to only 10 kg per car in 1960. By 1978, the figure had risen to about 60 kg, with most of it being used in the interior of the car. Legislation for passenger safety and the drive for longer mileage per gallon, in addition to other obvious reasons already discussed in the previous section, played a major role in increasing the use of plastics in cars. Polyurethane was increasingly used in the interior of cars for the production of seats and to cushion impact areas.

Legislation for increased fuel economy, following the oil crisis, has made it essential that cars operate more economically. The use of plastics gives an indirect reduction in energy consumption through the lower energy content of plastics parts used, which amounts to about 0.63 litre of oil equivalent for each pound of plastics. For a conservative estimate of an average use of 50 kg of plastics per car the energy saving for a year's total car production in the United States amounts to about 0.7 million TOE (Vowles, 1979). In car production, the change from metal or glass to plastics achieves a weight reduction, on a part-by-part basis, of 40–80 per cent, as shown in Table A10. This weight reduction offers an energy saving through the lower fuel consumption over the life of the car.

The combined energy saving in the production of plastics parts and the fuel consumption saving over a car-life cycle of 100,000 miles, for a conservative weight reduction of 40 per cent for the various plastics materials used in cars are presented in Table A11. This saving amounts to about 2.5–4 litres of petrol per pound of plastic used, which means that a car's fuel consumption could be reduced by 6–16 per cent and possibly more.

In the early 1980s about 0.85 million tonnes of plastics materials were consumed in the production of cars in the United States and in Western Europe the figure was nearer to 1 million tonnes. The amount of plastics used in each car varies from one manufacturer to another and depends on the model and size of the car.

The Italian (Fiat), French (Renault) and Japanese producers have a lead over other producers in the use of plastics materials, which amounts to about 60–80 kg/car or about 6–9 per cent of the total car weight. Recent models, such as the Panda and Strada in the Fiat series, contain a higher percentage of plastics, about 15 per cent of the outer car body. American car manufacturers use about 85 kg of plastics per car, but this accounts for just over 5 per cent of the total car weight since on average American cars are about two-thirds heavier than European cars.

The major plastics materials that are used in cars are polyurethane, ABS, PVC, HDPE and PP. Their main uses until now have been in the interior car parts but they are being increasingly used for exterior parts such as bumpers, cooling fans, fuel tanks, headlights and side mirror casings, window frames as well as add-on parts to improve the car aerodynamics. Of the major plastics, PP compound holds a promising future for increased use in exterior parts for cars.

Future plastics consumption per car would depend on the results of the developments in plastics materials processing techniques, improvements in composite materials' production, that would give

Table 1.8 Percentage composition of materials in average automobile

	1975	*1980*	*1990*
Plastics	3.5	5.7	9.2
Aluminium	2.9	6.3	11.9
Low-carbon steel	55.2	50.4	46.3
High-strength, low-alloy steel	6.0	6.5	7.9
Cast iron	16.2	13.6	7.9

Source: 'International Research and Technology, Rapra Trend', *Trade and Economic News Digest*, No. 68, February 1981.

them strength and durability, and, more importantly perhaps, on the level of energy price rises. By 1990, Western European cars, it is estimated, will be using between 100–150 kg of plastics materials, while American cars could use between 150–230 kg.

Car manufacturers emphasise the face that it is not simply the lightness and price of plastics that will determine their future role in cars, but more importantly, their performance. Aluminium is going to be the main competitor to plastics materials in this field, as it offers an equivalent weight saving. An estimate of what the distribution of materials used in car production would be in 1990 is given in Table 1.8: plastics could account for about 9 per cent and aluminium about 12 per cent of the total car weight by that year.

We believe more plastics could be used in the production of cars, and favour the more optimistic estimates of the average plastics consumption per car since we think that future increases in energy prices will promote this trend. Evidence from existing plastic-steel hybrid research cars in Italy and Japan suggests a labour saving of 5 hours in car assembly in addition to the considerable weight saving. These factors and the improvements that are already under way in producing and processing tailor-made plastics composites (glass- or carbon-reinforced fibres) will bring nearer the day when plastic-steel hybrid cars start coming on the market, perhaps before the end of this decade. The main problem facing the production of these cars on a mass scale is that of reorganising the existing lines of production.

The building industry, on the other hand, is the second largest market for plastics materials after packaging. However, plastics still form only a small percentage of the total materials used by the building industry. The construction of a new house in the United States, for instance, requires only one tonne of plastics materials (UNIDO, 1983), a small amount compared to the large volume of other materials involved. The industry as a whole consumed 3.5 million tonnes of plastics products in the United States in 1979.[12] The main uses of these materials were in pipes and fittings, floor covering, insulation, glues and resin bonding of woods and other various uses. The largest single product used in the building industry is PVC, mainly as pipes and ducts and to a lesser extent in floor covering, it accounts for about 40 per cent of all plastics used in the industry. Other important products used by the industry are HDPE, ABS and phenolic compounds. Plastics building materials help to save energy: they consume less energy to produce than their alternatives, they are light and hence will cost less to transport and to handle and, more importantly, they save energy when used for insulation. Plastics also offer corrosion resistance which is very useful in pipe applications. Also, competitive prices ensure continued growth prospects for plastics materials in the building industry. There is scope for increased use in pipes, insulation foams, door and window frames and many other applications, which are already expanding in Germany.

2 PRODUCTION RATIONALISATION AND OLIGOPOLY: A GLOBAL PERSPECTIVE

2.1. Trends in prices and production costs of petrochemical products

After the Second World War and throughout the 1960s, the role of oil and its derivatives as a feedstock source to the organic chemical industry was rapidly increasing at the expense of coal and tar products. This was partly due to the ease in handling and processing liquid feedstocks, but more importantly, a result of the availability of large quantities of oil and gas products from refineries at low prices since oil was cheaply priced by the monopolies of the oil companies which had a total control over its production, pricing and distribution. The cost of oil over this period was decreasing in real terms (Griffin and Teece, 1982); it was estimated that oil cost chemical manufacturers 50 per cent less than coal by the end of the 1960s.

Production costs of most petrochemical products were also declining over this period as larger automated chemical plants were coming on stream, exploiting economies of scale and effectively reducing the cost of production. As a result of the increase in the unit production capacity in the 1960s, producers were able to lower their unit investment costs by 25–65 per cent and their production costs by 14–55 per cent (see Table B1, Appendix B). The trend towards installing larger capacity plants continued into the 1970s with substantial decreases in unit capital investments costs as well as production costs.

The two factors of reduced investment costs and the decline in the price of oil in real terms during the 1950s to 1960s contributed to the fall in the wholesale prices of most commodity petrochemical products. For plastics and synthetic resins the drop was more than 15 per cent in the United States and about 6 per cent in Japan. The price of bulk petrochemical products fell more heavily: 35 per cent for ethylene and methanol and about 50 per cent for ammonia, ethylene oxide, polystyrene, styrene and vinyl chloride. For PVC and polyethylene the drop was 60–65 per cent. This was the average internal fall in prices for integrated firms in the United States and Western Europe in general.

By 1973–4 this pattern of declining prices for petrochemical products came to an abrupt end with prices rising sharply as a result of two factors:

1. the increase in the price of oil, the raw material and source of energy for petrochemicals; and
2. demand for petrochemical products outstripped supply when firms rushed to raise stocks at a time when supplies were inadequate and price levels were uncertain.

On average, 1974 product prices showed an increase of 200–350 per cent over the previous years. The prices started to stabilise by 1975 and the situation eased slightly. After 1976 demand started to pick up with prices rising slowly (see Table B2), reflecting an increase in demand and the higher cost of feedstock and energy components of these products.

Oil prices increased again in 1979–80, following supply shortages during the Iranian revolution and the subsequent Gulf war; as a result product prices rose, but more slowly than the producers had hoped since by this time the deep recession was world-wide with demand increasing at low rates. The oil price increase caused the energy component in the final product prices to become more significant, initiating a new pricing trend that brought production cost into conformity with the increased costs of energy, reflecting its long-term scarcity of supply in the form of gas and oil.

Prices of petrochemical products, especially commodity products, are strongly related to their production costs which include the cost of raw materials, energy and fuel, capital investment costs and wages and salaries. Prices of products tend to decline, in general, as the number of producers increases, with the technology becoming widespread and processes standardised, increasing the competition and forcing prices downward. Technological advances and process improvements also have an important corrective effect on prices.

Before the first increase in oil prices in 1973, the cost of raw materials accounted for a large proportion of the production cost of many petrochemical products, in particular for synthetic resins, plastics materials and fertilisers, where the cost of raw materials was

40–60 per cent of the production cost of these monomers and intermediate products.

This raw material cost-price relationship for these petrochemical products was further strengthened with the rise in oil prices. It was estimated that a 10 per cent rise in gas prices resulted in a price increase of 3–4 per cent for ethane and 5–6 per cent for ammonia. For intermediate products a 10 per cent rise in ethylene prices resulted in an increase in the production cost of PVC in the order of 5–7 per cent, about 5 per cent for polyethylene and 6–7 per cent for polystyrene when produced in plants of a comparable capacity.[13]

The rise in oil and gas prices meant that raw material costs will account for a larger share of the production cost than before, making future product prices more sensitive to and in line with the prices of energy sources in general.

2.2. The changing structure of production costs

The production costs of petrochemical products vary greatly, depending on the type of product, the process of production and raw materials used, the plant size, its location and its investment costs. Also, wages and productivity of the manpower employed contribute, albeit as a very small share, to the cost of production. The importance of each of these factors in relation to the production cost of final petrochemical products is assessed below.

The impact of oil price rises on the production cost depends on the energy intensiveness of these products and how far downstream they are processed. Table B3 gives the structure of the production cost for some petrochemical products produced by medium and small size plants in Western Europe as of 1977. These costs may be slightly different from those prevailing in the United States or Japan and, while it is obvious that for larger plants the production costs would be effectively reduced by at least 10 per cent on average from those shown in this table. Their significance, however, lies in showing the importance of raw materials and energy costs in 1977 relative to the situation in previous years. They also illustrate how these costs differ from one product to another, depending on its energy content and the complexity of its production process.

Products such as methanol, vinyl chloride, PVC, styrene, LDPE

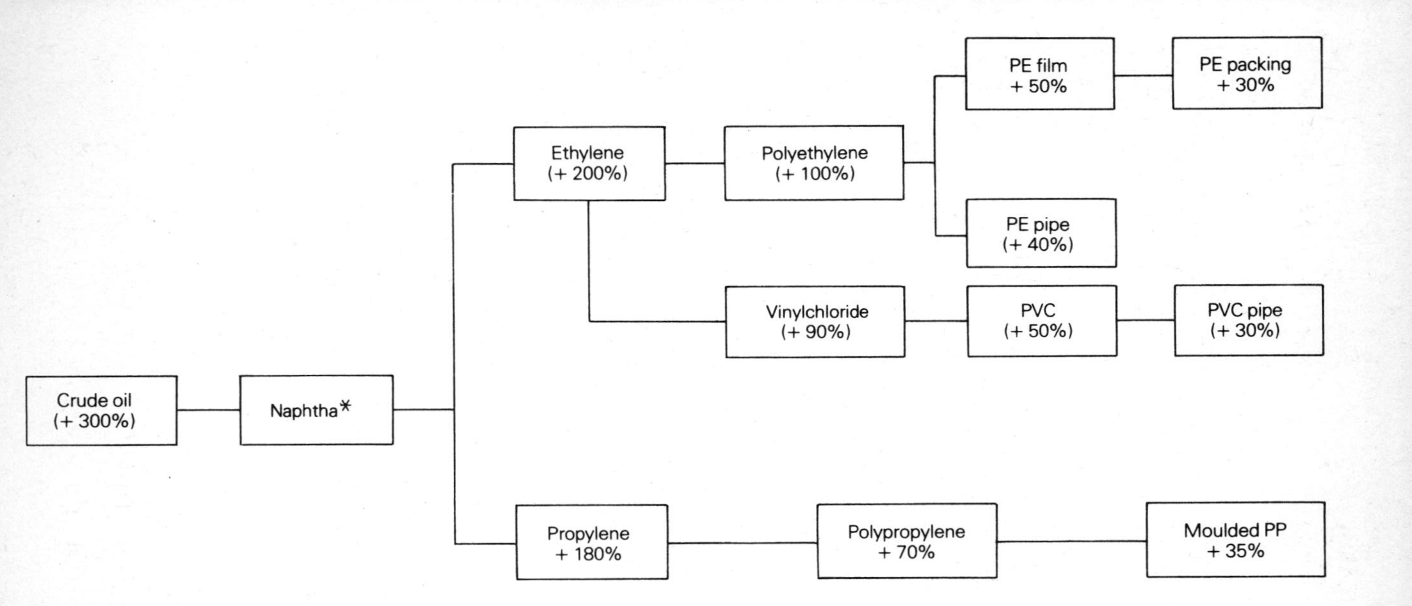

* Naphtha price depends on the offer and demand situation: it can vary to some extent, independently from the crude price. During the considered period, the naphtha price increases amounted to 400 per cent.

Figure 2.1 Impact of the crude oil cost increases on petrochemical products cost between June 1973 and June 1974

Source: 'Plastics after the oil crisis: the ICI view', *Plastics Today*, December 1974, p. 14.

Table 2.1 Increase in olefins manufacturing cost

	Naphtha steam cracking	*Naphtha steam cracking*	*Naphtha steam cracking*
Capacity tonnes/year ethylene	300,000	300,000	300,000
Economic conditions	Prevailing in 1972	Prevailing in 1977	Prevailing in 1977 Unit erected in 1972. Investment in 1972
Fixed capital cost US$ m	104	184.3	104
Manufacturing cost US$ thousand			
Raw materials	21,150 (42.2)*	129,600 (71.6)	129,600 (78)
Utilities	1,080	2,200	2,200
Catalysts and chemicals	820	1,100	1,000
Manpower	500	1,100	1,100
Other charges	6,750	12,000	12,000
Amortization and return	19,800 (39.5)	35,000 (19.3)	19,800 (12)
Total manufacturing cost	50,100 (100)	180,900 (100)	165,700 (100)

* ( ) : percentage values introduced by authors

Source: UNIDO, ICIS. 83, December 1978.

and HDPE have production costs that are dominated by the raw materials costs. This dominance decreases for nylon and acrylic fibres, polyesters, ethylene glycol and coprolactam, whose raw materials costs contribute less than 40 per cent to the total production cost. This relationship between raw materials and total production costs is very strong for basic and intermediate products such as synthetic resin monomers and fertilisers whose energy content is high and process of production very capital intensive and large in scale.

The effect of raw materials costs decreases for products further down stream, whose production involves extensive processing before becoming ready for use as final products. Figure 2.1 shows this relationship between the rise in raw materials prices and the resulting production costs increases of the downstream products.

After 1974 feedstock and capital costs increased relative to other factors of production. This can be illustrated by analysing the changing production costs of ethylene, the most basic of petrochemical products, over the period 1972–7. Table 2.1 shows that for a 300,000 tonne per year ethylene cracker, that came on stream in 1977, the cost of raw materials was 6.13 times that of an equivalent one operating in 1972, while capital costs increased by 1.76 times only. The overall effect was that by 1977 total production costs had grown by 3.6 times over those prevailing in 1972 for a similar plant. In 1977, raw materials for ethylene production had become the dominant component of production costs, rising from 42.2 per cent in 1972 to 71.6 per cent in 1977 (Table 2.1). On the other hand, although capital investment costs rose over this period, their share of the total production costs had declined in relative terms from 39.5 to 19.3 per cent.

The last column of Table 2.1 illustrates another important feature of the production costs structure of ethylene; where a plant built in 1972 showed a lower total production cost than an equivalent plant built in 1977, due to its favourable investment costs, this allowed it to achieve higher profits or to compete more successfully with plants of the same size built in 1977.

2.3. Plant capacity and its impact on production costs

The other important factor affecting the production costs of ethylene is that of the plant size, whose significance has also changed as a result of oil price rises. In 1972 doubling the size of a 150,000 tonne per year ethylene plant would lead to a 14 per cent reduction in production costs (which include ROI but not by-products benefits), with the capital investment share of total production costs declining from 43.6 per cent for the 150,000 tonne per year plant to 39.5 per cent for the 300,000 tonne per year plant due to the non-proportional relationship that exists between investment costs and the plant size, which is characteristic of petrochemical plants and refineries, and has the following form:

$$I_A = I_B (C_A / C_B)^n$$

where I = investment costs
C = plant capacity

n = power coefficient whose value ranges between 0.6 and 0.8

A, B = two plants, identical but for their capacities.

This reduction in manufacturing costs is reflected in the prices of produced products, where for ethylene a 17 per cent reduction in the price per tonne occurs when a larger capacity plant is used (see Table 2.2).

Table 2.2 Steam cracking economics (European conditions)

	Conditions prevailing in 1972		*Conditions prevailing in 1977*	
	Naphtha steam cracking	*Naphtha steam cracking*	*Naphtha steam cracking*	*Naphtha steam cracking*
Capacity tonne/year ethylene	300,000	150,000	300,000	150,000
Fixed capital cost US$ m	104	67	184.3	118
Production cost US$ thousand				
Raw materials	21,150 (42.2)*	10,570 (36.1)	129,600 (71.6)	64,800 (66.4)
Utilities	1,080	540	2,200	1,100
Catalysts-chemicals	620	310	1,000	500
Manpower	700	700	1,100	1,100
Other charges	6,750	4,350	12,000	7,700
Amortization and return	19,800 (39.5)	12,750 (43.6)	35,000 (19.3)	22,400 (22.9)
Total	50,100 (100)	29,220 (100)	180,900 (100)	97,600 (100)
Products prices $/tonne				
Ethylene	90	108	320	340
Propylene	55	66	220	230
Butadiene	150	186	370	390
LPG	32	32	130	130
Gasoline	45	45	168	158

* ( ) Percentage values introduced by authors

Source: as in Table 2.1.

In 1977, doubling the size of the plant would result in a production cost reduction of 7 per cent, only reflecting the declining share of the capital investment cost component relative to that of the feedstock, which accounts for 66–71 per cent of total production cost, depending on the size of the plant. This implied that, as production costs become less sensitive to capital costs, small market countries, particularly developing countries, that could control their feedstock prices were still able to produce ethylene or other commodity petrochemical products without the need to invest in very large-capacity plants. Price rises of raw materials and fuel had an increasing impact on the costs of chemical equipment and plants, which more than doubled over the period 1970–9, with annual increases higher than the rate of inflation. Construction costs are likely to be rising at least at the same rate as the consumer price index, for the rest of the 1980s.[14] Other factors which lead to higher production costs include the costs of additional equipment and devices or modified designs to reduce the effects of pollution and improve safety.

The effects of higher wages are, however, less pronounced than other factors of production due to the low labour costs per unit of production in this capital intensive industry, especially in the larger capacity and highly automated plants. In the last two decades this phenomenon had been exemplified by rising wages in this branch of the chemical industry, while product prices were falling until 1973.

Up to the 1970s, technological improvements and scientific advances were the major contributors to lower product prices, but since the early 1970s there has been a decline in the number of major advances and new products being developed. It is also unlikely that major changes in existing production processes will take place in the near future which will revolutionise the petrochemical industry, as did the large steam crackers and the development of polymerisation techniques. R & D expenditure has declined and also it is difficult to change the existing structure of the industry, where new processes and new plants have to compete with older plants over which they lack any appreciable advantages due to their higher investment costs and to replace the existing equipment of the affected sectors of the industry quickly. Changes would have to take place slowly over long periods, particularly when there is overcapacity and demand is weak. It is for these reasons that new products such as LLDPE, for instance, are facing difficulty in penetrating the markets as vigorously as similar products in the 1960s.

Present emphasis in the industry is more on improving existing processes and building larger capacity plants that are flexible in their use of a variety of feedstocks. Also, another current interest is the exploitation of low-cost raw materials such as methanol and SNG. However, these developments are not expected to make a major impact on the industry in the near future unless there is a major technological breakthrough.

Plant size has a direct effect on the investment costs for petrochemical products and invariably on their cost of production and selling prices. This effect, however, varies, depending on the type of product, its level of *maturity*, its production cost structure and the complexity of the production process. The range of plant sizes or capacities used also varies, depending on the process of production and the extent of demand for these products.

Table B4 illustrates this relationship for a number of plants operating in the United States, where conditions can be considered as similar to those of other industrialised countries in general. It is apparent that investment costs per unit output of product vary considerably from one product to another, with PET having the highest investment cost requirement of $2,150 per tonne/year (for the large capacity plants), whereas polystyrene requires only $855 per tonne/year. Investment costs for PVC at $1,514 per tonne/year are also higher than those of other polyolefin products such as HDPE or LDPE. LLDPE, due to its simpler and improved low pressure production process, requires an investment per unit output that is 30 per cent lower than that for LDPE.

The impact of plant size on production costs is given in Table B5 for a range of plant sizes – large, medium and small – whose capacity ratios are in the order 4:2:1. The transfer prices of the products shown in the table include production costs with an allowance of 25 per cent return on investment (ROI), assuming plants operate at a load factor of 85 per cent. The effect of using large-size plants on the transfer price is found to be an overall reduction in the range of 7–16 per cent for most products, which are arranged in four groups according to the percentage reduction in price resulting from lower production costs and are shown graphically in Figure 2.2.

The above results show that some products are more sensitive to economies of scale than others, with the implication that two factors

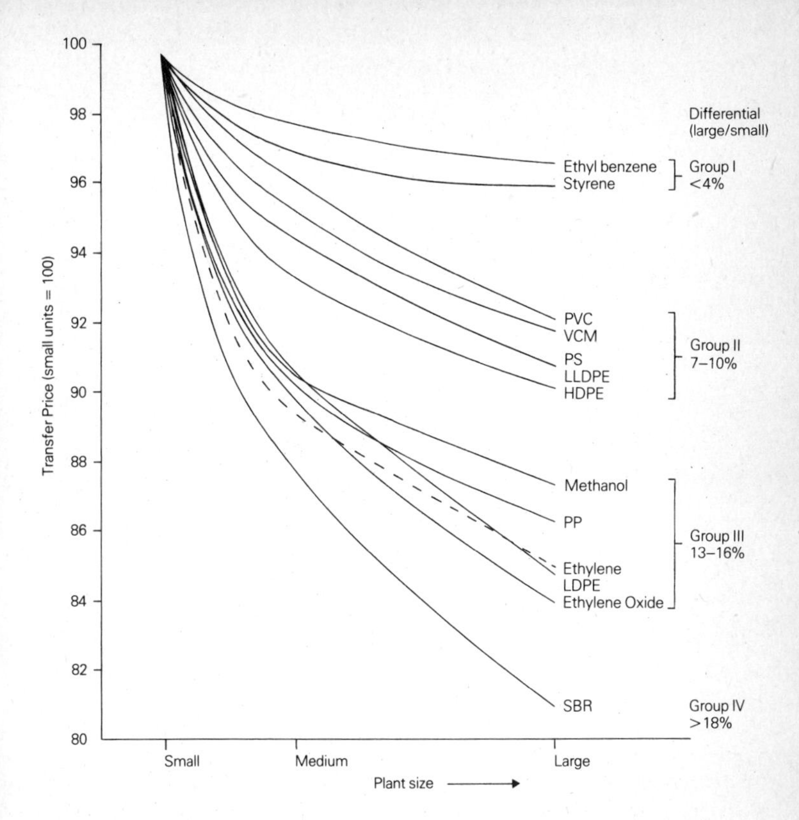

Figure 2.2 Economies of scale for selected petrochemicals

Source: UNIDO, ID/WG. 336/2, May 1982.

must be taken into consideration when evaluating what plant size should be chosen for which product: First, the significance of net reduction in costs of production and transfer prices. Second, the percentage reduction in investment costs per unit output and the overall investment requirements for large plants and the compatability of the production capacity with the size of the market. The implications of these two factors are that for products that are less sensitive to economies of scale and whose investment costs per unit of output are high, such as LDPE, it would be more appropriate for smaller producers (for example, developing countries with small home markets) to employ small- or medium-size plants and yet still

Table 2.3 Cost of naphtha-based ethylene in Europe (Capacity: 50,000 tonne/year current $)

	1972	*1977*		*1980*		
		1972 unit	*New unit*	*1972 unit*	*1977 unit*	*New unit*
Fixed capital cost, US$ m	43.7	43.7	77.5	43.7	77.5	89.8
Manufacturing cost, US$/tonne:						
Raw materials	70.5	430.0	430.0	1,033.4	1,033.4	1,033.4
Bi-product feedstock	−79.9	−283.0	−283.0	−625.7	−625.7	−625.7
Utilities	3.6	7.3	7.3	6.8	6.8	6.8
Catalyst and chemicals	2.1	3.3	3.3	1.7	1.7	1.7
Manpower	2.3	3.7	3.7	23.0	23.0	23.0
Other charges	22.5	40.0	40.0	100.0	100.0	100.0
Depreciation	87.4	87.4	155.0	87.4	155.0	179.7
Total production cost	108.9	288.7	356.3	626.6	694.2	718.9
Raw materials/ depreciation	0.8	4.9	2.8	11.8	6.6	5.7
Depreciation/total production cost (%)	80	30	43	14	22	25

Source: UNIDO, ID/WG. 336/3, May 1981, p. 187.

maintain a competitive production capability by operating these plants at near full capacity.

2.4. The effects of 'age' of plants on production costs

A factor superimposed on, and influencing the production cost structure of petrochemical products, relates to the combination of the age and size of plants. Tables 2.3 and 2.4 illustrate the effects of this factor on small size (50,000 tonnes/year) and large size (300,000 tonnes/year) ethylene crackers, where the production

Table 2.4 Cost of naphtha-based ethylene in Europe (Capacity: 300,000 tonne/year current $)

	1972	*1977*		*1980*		
		1972 unit	*New unit*	*1972 unit*	*1977 unit*	*New unit*
Fixed capital cost, US$ m	153.3	153.3	271.6	153.3	271.6	329.2
Manufacturing cost, US$/tonne:						
Raw materials	70.5	430.0	430.0	1,033.4	1,033.4	1,033.4
Bi-product feedstock	−79.9	−283.0	−283.0	−625.7	−625.7	−625.7
Utilities	3.6	7.3	7.3	6.8	6.8	6.8
Catalyst and chemicals	2.1	3.3	3.3	1.7	1.7	1.7
Manpower	2.3	3.7	3.7	23.0	23.0	23.0
Other charges	22.5	40.0	40.0	75.5	75.5	75.5
Depreciation	51.0	51.0	90.5	51.0	90.5	109.7
Total production cost	72.2	252.3	291.8	563.7	603.2	621.9
Raw materials/ depreciation	1.4	8.4	4.7	20.2	11.4	9.4
Depreciation/total production cost (%)	70	20	31	9	15	17.5

Source: UNIDO, ID/WG. 336/3, May 1981, p. 148.

cost structures of plants of different ages are compared under conditions prevailing in Western Europe in 1977, 1980.

Again, these tables show clearly that due to their lower fixed investment requirements plants built in 1972 had an obvious advantage in terms of their production costs over similar plants that came on stream in 1977 and 1980. In 1972, depreciation charges accounted for 80 per cent of the total production costs (excluding ROI and interest on working capital), a trend which was reversed in 1977 when feedstock charges became the dominant factor of production, as can be seen from the feedstock/depreciation ratio for a 1972 plant, rising from 0.8 in 1972 to 4.9 in 1977 and 11.8 in

1980 respectively. The trend was similar for plants that were coming on stream later on.

A new unit coming on stream in 1977 had a feedstock/depreciation ratio of 2.8, whereas a new unit coming on stream in 1980 had a ratio of 5.7; in other words, feedstock costs were 5.7 times those of depreciation, showing that, although capital investment costs were rising in absolute terms, the second oil price rise had appreciably increased the significance of feedstock costs over those of depreciation. Thus, by 1980, the advantage of older plants over the new units in terms of production costs, had been effectively reduced from the 1977 level. This is shown by the differences in production costs and by the similar feedstock/depreciation ratios of 6.6 and 5.7 for the 1977 and 1980 units respectively.

Depreciation charges accounted for 43 per cent of the total production costs for a new unit that came on stream in 1977, which had declined to 25 per cent for the new units which were coming on stream in 1980. For the 1974 and 1977 units, operating in 1980, the depreciation costs accounted for 14 per cent and 22 per cent of total production costs respectively, showing that the influence of capital investment charges over production costs had been greatly reduced and the advantage held by the older plants in this respect diminished.

The large capacity plants show a similar trend, but they have higher raw materials/depreciation ratios than the small unit plants, reflecting the prominence of feedstock costs. Also, the contribution of depreciation to total production costs had declined, and in 1980 were 9, 15 and 17.5 per cent for the 1972, 1977 and 1980 units respectively (cf. small plants), showing the advantages of economies of scale.

The tables also show that small-size plants built in 1972 had a production cost of $626/tonne in 1980, which enabled them to compete with new large-capacity plants that were coming on stream in that same year, with production costs of $622/tonne. However small plants built in 1977 or later would not be able to compete with the larger capacity plants of the same generation since their production costs were about $700/tonne each compared with $603/tonne and $622/tonne for the large capacity units.

From this analysis we can conclude that older plants have a production cost advantage over new plants of the same size. However, Tables 2.3 and 2.4 ignore the fact that production cost structure of new plants may be favoured by technological improvements in design, processing, energy saving, new catalysts and feedstock flexibility, all of which affect overridingly the production yield and the economics of the new plants.

Production costs are also significantly affected by *location*, which is expressed as a quick conversion factor that relates the cost of fixed capital investments in various regions of the world to those prevailing in the Gulf coast of the United States, that is taken as a reference base whose value is unity. Location factors reflect the state of development of the country's infrastructure, its labour productivity and wage rates in the construction sector (mechanical and civil engineering), its technical backup and engineering services. These factors vary from one country to another, with Japan having a lower factor (0.9) than the United States, and Indonesia having a factor more than twice that of the United States. High location factors may offset the advantages of low cost feedstock in oil-producing countries since they raise the fixed capital costs and return on investment rate falls to maintain competitiveness. A more detailed analysis of the effects of location factors on the production costs of a number of products and the competitive position of developing oil-producing countries is given below (see Section 4.3 and Table D3 in Appendix D).

2.5. The process of rationalising petrochemical production

The situation which developed after 1973 had profoundly altered the production cost structure of the major petrochemical products and started to change the roles of the major operators in this industry. In particular the traditional roles of the major chemical companies and of the chemical subsidiaries of the major oil companies were experiencing a phase of slow, but steady, change. The major chemical companies became anxious about the security of their raw materials and feedstocks and at the same time they were facing increased competition from the chemical subsidiaries of the oil majors. The oil companies were trying to diversify their operations and benefit from the new situation, in which the role of feedstock as an important production cost component had become more prominent, thus further reducing the rates of return on investment from the levels realised by the petrochemical producers during the 1960s.

This aspect of the industry is discussed below for the three main producing regions of petrochemicals: the United States, Western Europe and Japan. The roles of the major chemical and oil companies and the changing pattern of production is analysed and the degree of product capacity concentration and the extent of vertical integration by the major producers is also discussed. Finally, the speed with which rival firms were able to generate imitative process technologies without the need to license from the patent holder and the resulting impact on the process of rationalising the petrochemical industry is briefly covered.

2.5.1. The implications for the major chemical companies: the feedstock position

By 1975, chemical companies were looking for ways of securing their feedstock needs and were re-evaluating their relationship with their oil and feedstock suppliers. This was particularly necessary for producers in Western Europe and Japan where industries have been based on the use of naphtha feedstock, and are expected to remain so until the late 1990s.

Chemical companies were faced with a difficult situation: oil prices had increased, thus increasing their costs of production, and naphtha prices were increasing at faster rates than oil prices during 1979–80; some chemical producers blamed oil refiners for this situation claiming that, unable to market the heavier products (such as gas oil), they compensated by raising naphtha prices in relation to production costs instead of market value.[15] Further, in Western Europe and Japan, due to its dependence on naphtha, the petrochemical industry has to compete with the automotive industry for its supply of feedstocks in a tight market, since oil refining in these two regions is weighted in favour of the middle distillates and fuel oil for heating and other industrial uses (Table B6). The industry in many cases was forced to import its additional needs of naphtha by purchases on the international spot market.

This situation made the major chemical companies uneasy about their feedstock security, particularly in Western Europe and to a lesser extent in Japan where the import and distribution of oil is carried out with a high degree of government control and backing. In brief, the major chemical companies had become more vulnerable in their relationship with their feedstock suppliers, the oil companies, which had already established a foothold in the petrochemical market via their chemical subsidiaries. In the United States the situation was easier, because a higher fraction of light distillates is obtained per barrel to satisfy the large demand for petrol, because the industry relies on natural gas for olefin production and, further, because oil and gas prices have not so far been fully decontrolled.

To obtain some independence from the oil companies, the chemical companies reacted in three ways:

1. By investing in upstream operations, either indirectly through the acquisition of large shares in small oil companies or directly by building their own refineries or going into joint ventures with oil companies in new large-scale chemical refineries or catalytic ethylene plants.
2. By concluding direct long-term contracts with oil-producing countries or by investing in joint ventures in these countries in return for oil entitlements.
3. By increasing research into processes based on alternative feedstocks such as methanol and SNG with the long-term objective of moving away from light-grade oil-based materials.

In West Germany, BASF and Hüls had connections with the oil companies. Bayer and BP together formed Erdölchemie, a 50/50 joint venture, and BASF and Shell formed another joint petrochemical company, ROW. Only Hoechst, of the major chemical companies, remained non-integrated and preferred to obtain its feedstock and ethylene needs from its traditional suppliers.

In the United Kingdom, on the other hand, ICI had acquired some refining capacity to meet some of its energy needs. It also moved to form a joint venture with BP in a large ethylene cracker at Teeside.

The chemical producers of France and Italy had some degree of feedstock security since the oil sector in both countries was government controlled. The two major oil companies in France, Total and Elf-Aquitaine, had 70 per cent of their shares controlled by the French government, and the ENI oil company in Italy was totally government-owned. Some of the largest chemical firms in Italy and France, such as ANIC, a subsidiary of ENI, and Ato-Chimie and CdF-Chimie, also had a large degree of government control. These companies were therefore in a more secure position than the privately owned companies, such as SIR and Montedison (Italy) or Rhône-Poulenc and Produits Chimiques (France).

These independent chemical companies have recently started to take appropriate measures to secure their feedstock supplies by forming joint ventures with government-owned chemical companies or through long-term oil supply contracts with the government-owned oil companies. In Italy, Montedison and SIR established their own refineries to meet their requirements of naphtha, buying oil on the international market, but at the same time they obtained joint control of ethylene crackers with the largest oil processor in the country, ENI.

The situation in Japan was, as mentioned above, less serious than in Western Europe due to the government involvement in importing and distributing oil supplies, but the country as a whole was at a severe disadvantage due to its lack of hydrocarbon resources; this was a major concern for the Japanese government which encouraged the Japanese petrochemical producers to establish direct contacts and joint ventures with oil producers of South-East Asia and the Middle East.

In the United States, Dow Chemicals and Monsanto were among the leading chemical firms that have adopted backward integration since 1975 by investing in oil refineries and obtaining some oil reserves. (Dow owns a 200,000 barrel per day oil refinery in Freeport, Texas, while Monsanto has a 32,000 barrel per day refinery, jointly owned with Conoco.) Du Pont's recent acquisition of the Conoco oil company is a major step towards securing its feedstock supplies. Also, Du Pont had a ten-year agreement with Shell Chemical to obtain one-third of its olefin requirements.

2.5.2. The implications for the major oil companies: down-stream investments

Major oil companies in Western Europe and the United States had had substantial petrochemical operations even before the oil price rises. However, after 1973 the major oil companies had an opportunity to participate more effectively in the chemical markets by virtue of their control over oil and, to a lesser extent, over gas supplies. The high cash flows and profits after each oil price rise, the access to and control of raw materials gave the oil companies opportunities to invest further in down-stream operations where conditions have become more favourable to them than to the traditional chemical companies.

By the late 1970s, the wave of nationalisation of the interests of the multinational oil companies in most oil-producing countries, especially the OPEC members, which gained momentum after the first oil price rise of 1973, had been completed. The companies role as oil producers being greatly reduced, and they became, effectively, processors and distributors of oil and gas products. The trend to diversification down-stream was therefore further strengthened.

The involvement of the oil companies[16] in the petrochemical industry is clear from their historical capital expenditure data (Table B7) which show increasing investment in petrochemical plants following the oil price rises. The largest investments were in 1976 and later and were mainly concentrated in the United States and Western Europe, which together accounted for about 60 per cent of the world total. In the United States, investment in petrochemical plants exceeded those in refineries over the period 1976–8, whereas in the Middle East substantial investments in chemical plants and natural gas liquid plants came into effect in 1977 and, by the end of 1980, totalled $1.75 billion in chemical plants and $2.5 billion in natural gas liquid plants. It is significant that investments in chemical plants in the Middle East had by 1980 become equivalent to those of Western Europe. In Africa (mainly the oil producing North African states and Nigeria), the pattern of increased investments in chemical and natural gas liquefaction plants was similar to that of the Middle East.

The data in Table B7 also reveal that the trend of the early 1970s, when investments in chemical plants in Western Europe and the United States accounted for more than 80 per cent of the world total in this sector by oil companies, was changed by the later 1970s with increased investment in oil-producing countries.

World-wide investments in oil-processing plants (oil-refining, petrochemical, and natural gas liquefaction) amounted by the end of 1980 to $27.5 billion, which is the equivalent of 26 per cent of the major oil companies' total capital and exploration expenditure. Their sales of petrochemicals, which in 1980 totalled $38.8 billion, although representing only 6.6 per cent of their total revenue, shows the increasing importance of the group of companies as petrochemical producers.

Chemical companies, in their turn, were seeking security for their feedstock supplies, following the oil price rises. The increasing size and cost of the new petrochemical plants encouraged the two groups of companies to cooperate more closely and form joint ventures to minimise risks and to ensure markets and supplies in various parts of the world. These were more concentrated in Western Europe, where the following projects involving large-scale ethylene crackers were completed with the direct cooperation between major oil and chemical companies: the ICI/BP Unit in the United Kingdom; the BASF/Shell Unit in West Germany; the Bayer/BP Unit in West Germany; and the CFP/ELF Unit in France.

In other cases, long-term contracts were concluded between the major oil and chemical companies, whereby the former supply the latter with major petrochemicals on the basis of contracts that could extend over periods of 5 to 10 years. In West Germany, the major oil operators Caltex, Marathon, Exxon and Gulf sell over 50 per cent of their ethylene output on this basis to the three major chemical sisters. Similar arrangements were made in the United States for the supply of major petrochemical products. But joint ventures between the two groups of companies were much less significant in the United States than in Western Europe.

2.6. The roles of the major oil and chemical companies in the process of restructuring the petrochemical industry

2.6.1. The Western European situation

The period since 1976 has been marked by the increased participation of major oil-producing companies in down-stream operations, covering basic and intermediate petrochemical products as well as the major plastics materials. In 1980, the oil companies operating in Western Europe controlled between 40–60 per cent of the basic petrochemical production (see Table 2.5) as a result of their active intervention in the industry, following the oil price rises.

In intermediate and plastics products, the oil companies controlled between 20–30 per cent of the total production capacity. However, the rates of growth of the major oil companies shares in these products since 1976 were very high, as shown in Table 2.6. It is important to note that three companies were very active in this

Table 2.5 Major oil companies' share in basic petrochemicals

Basic products	*Major oil companies share in 1980 (%)*
Ethylene	43.5
Butadiene	61.3
Benzene	43.0

Source: Based on Table B8, Appendix B.

Table 2.6 The percentage shares of major oil companies in petrochemical capacity (Western Europe)

Product	*1976*	*1982*	*Growth*
Ethylene oxide	20.8	30.4	46.1
Styrene	32.4	33.3	2.7
LDPE	30.1	32.6	8.3
HDPE	21.9	25.1	14.5
PVC	15.3	21.3	40.0
PP	22.3	27.8	20.0

Source: Based on Table B8, Appendix B.

area, namely, Shell, BP and Exxon. Other active participants in this process of restructuring were the independent petrochemical producers including government-backed companies and petrochemical producers whose main activity is not in chemicals. This group of companies had been rapidly increasing its share of the petrochemical capacity since 1976 which, by 1982, had grown to between 16 per cent and over 30 per cent for various products. On the other hand, the chemical companies' share of the major petrochemical products capacities had declined in favour of the increased participation of the oil companies and independent producers in Western Europe, as shown in Table 2.7.

The restructuring process was well under way in Western Europe in the early 1980's. Some companies cut their excess

Table 2.7 The changing shares of major chemical companies in petrochemical capacity (Western Europe)

Product	*1976*	*1982/3*	*Growth*
Ethylene	33.5	33.2	0
Benzene	26.3	27.4	4.1
LDPE	48.1	34.5	−28.3
HDPE	62.5	50.9	−18.5
PVC	62.8	54.6	−11.1
PP	71.5	55.7	−22.1
Styrene	51.1	47.6	−6.8
Ethylene oxide	67.2	47.0	−30.1

Source: Based on Table B8, Appendix B.

capacities closing down older plants in an attempt to improve their financial situation at a time when demand for plastics was low due to the recession, and also in expectation that in the future more plants coming on stream would be owned by the major oil companies. The following developments that were taking place in Western Europe are indicative of this process. In the United Kingdom, BP swapped its PVC capacity with ICI's LDPE and closed down its remaining older PVC plants. In France, Rhône-Poulenc pulled out of plastics production and its HDPE and PP shares were acquired by BP, while its PVC capacity was taken over by Elf and CFP of France. Union Carbide and Monsanto were pulling out altogether from the European market after selling their shares in LDPE, ethylene oxide and thermosetting products to BP. Western European major chemical companies were also increasing their activities in the United States petrochemical market, particularly in the following products: ethylene (ICI and Solvay), HDPE (Hoechst and Solvay), ethylene glycol (BASF) and polyethers (Bayer). At the same time, some major chemical companies were modifying their intentions for future bulk petrochemical production. Objectives of 50/50 or even 60/40 combinations of speciality chemicals and bulk petrochemicals production were declared more than once (Bennett, 1982).

Table 2.8 The percentage share of major oil companies in petrochemical capacity (United States)

Product	*1976*	*1982/3*	*Growth*
Ethylene	42.5	53.8	26.6
LDPE	22.5	27.6	22.7
HDPE	21.9	32.6	48.9
PP	47.4	52.2	10.1

Source: Based on Table B9, Appendix B.

Table 2.9 The percentage share of major chemical companies in petrochemical capacity (United States)

Product	*1976*	*1982/3*	*Growth*
Ethylene	40.0	32.9	−17.7
LDPE	40.4	40.1	−0.7
HDPE	44.4	37.5	−15.5
PP	38.4	31.7	−17.5

Source: Based on Table B9, Appendix B.

2.6.2. The situation in the United States

The restructuring process in the United States did not differ very much from that of Western Europe, except that overcapacity was not a major problem in the United States and market prospects were better. The major American oil companies had the highest increases in production capacities in the three major plastics, LDPE, HDPE and PP. The leading participants in these increases were Shell, Exxon, Gulf and Mobil. On the other hand, the major chemical companies position had declined, although the leading companies, Union Carbide, Dow and Hercules, were still adding new capacities, their share of the whole American market relative to what it had been in 1976 was declining in favour of the major oil companies and the independent companies, as shown in Table 2.9. Tables B8 and B9 summarise the changing structures of the petrochemical capacity ownership in Western Europe and the United States respectively.

The ownership of production capacities of two products, PP and HDPE, are strikingly different from the rest of olefin products. In the United States, the oil companies hold over 50 per cent for PP and 37 per cent for HDPE, but only 28 per cent for PP and 24 per cent for HDPE in Western Europe. This situation is attributed to the feedstock base of the petrochemical producers. In the United States oil companies hold a large share of the PP market because of the availability of refinery off-gas to them, while major chemical companies are normally ethane-based olefin producers. On the other hand, olefin production in Western Europe is naphtha-based which yields propylene and other off-gases, allowing the major chemical companies to keep a large share of the PP and HDPE market.

Finally, it should be noted that a large share of the new ethylene and LDPE plants coming on stream in the United States, Canada and the EEC countries since 1977 were accounted for by the *seven sisters* with new capacities of 40 per cent for ethylene and 83 per cent for LDPE, with Shell and Exxon playing the leading role in these investments.

The process of restructuring the petrochemical industry and the measures taken by its main operators were not restricted to their local markets but have become part of a world-wide process which has taken two directions:

1. The participation of those operators in joint ventures with the oil-producing countries, leading to the formation of large consortia, operating in Saudi Arabia, Iran and Qatar in the Middle East, Alaska and Canada, in North America, and in Australia. The operators involved in these consortia include: oil companies such as Exxon, Shell, Mobil and Gulf; chemical companies such as Dow and Mitsubishi and government-backed independents such as CDF-Chimie (France) and Anic (Italy). These moves aim to secure supplies of feedstocks and obtain access to cheap oil and gas resources.
2. A number of oil companies (Gulf) and chemical companies (Monsanto and Union Carbide) based in the United States abandoned their small market shares in Western Europe and moved instead into the American market. The European

> chemical and oil companies also played a dominant role in foreign investments in petrochemicals in the United States, which increased after 1973 (see Table 2.10), the main target being plastics materials.

The growing saturation of the European market with many products, particularly fibres and plastics, prompted the operators to shift their attention to the American market which has a vast industrial structure that can absorb the production of a substantial number of large-scale plants. Petrochemical producers moved into the United States mainly to be close to a large market with a high growth potential, especially in packaging, where the paper and paperboard markets had not been penetrated to the extent achieved in Western Europe, and in the car industry for the increased use of automotive plastics parts.

2.7. The oligopolistic structure of the petrochemical industry

It is characteristic of the petrochemical industry that a small number of companies controls the production facilities, the technology and know-how and also has a hold on feedstocks and raw materials supplies.

The largest petrochemical producers are also characterised by their extensive degree of integration, covering up-stream operations such as oil refining and ethylene production as well as down-stream operations covering wide areas of final production, such as plastic processing and marketing. This degree of integration from oil production and processing further down to marketing of final products ensures for these companies a relatively stable market share and a sound base for future growth and increasing exports. It also allows them to capture more profits at every stage of the production process and to use transfer pricing.

Ethylene is among the most integrated products. The degree of integration in ethylene derivatives is much higher in Western Europe than elsewhere due to the competition between rival firms on the European market and because of the large-scale operations of naphtha-based plants, controlled by the large firms.[17]

Only small amounts of ethylene originating from Mexico, South Korea and Algeria are traded world-wide, but substantial inter-EEC

Table 2.10 Foreign investment in the chemical industry

Year	*Total foreign investment* $ b	*European share of total* %
1973	2.9	71
1975	3.7	72
1977	4.8	77
1979	7.1	74

Source: Du Pont, Society of the Chemical Industry, 1981.

trade in ethylene takes place as a result of the established ethylene grid which serves to facilitate easy supply between various users whenever imbalances in production occur. A similar arrangement also exists in the Gulf coast of the United States, where most of the ethylene is also used in a captive market.

Propylene, cumene, paraxylene, DMT, TPA and acrylonitrile production is heavily integrated where companies, producing these materials, use them captively to produce the end products.

Benzene and xylene producers also have a considerable degree of back-integration into refining, but still lower than the level attained by olefin producers. Most aromatic producers are also involved in olefin production in Western Europe and the United States, with the exception of two companies, Mobil and Cepsa of Spain.

In Japan, however, only a limited degree of integration exists between refiners, olefins' and aromatics' producers. Only two groups of companies are fully integrated, Mitsubishi and Tonen petrochemicals.

Aromatics are traded more widely, particularly benzene and xylenes, than olefin products. But propylene and butadiene are traded in significant quantities, with the United States and Western Europe being net importers of the two products.

The degree of integration of refining, olefin and aromatic production of the major Western European and Japanese petrochemical producers is shown in Tables B10 and B11.

2.7.1. Product specialisation by leading firms

In the United States, the largest four producers for each of the major

Table 2.11 Petrochemicals production capacity concentration in the United States, 1980

Product	*Largest four producers (%)*	*Largest eight producers (%)*
LDPE	48.5	80.0
HDPE	49.0	81.0
PP	57.0	86.0
PVC	44.0	71.0
PS	52.0	79.0
ABS	90.0	100.0 (largest 6)

Source: Based on data obtained from *Modern Plastics*, January 1980.

Table 2.12 Capacity of the largest petrochemical producer (1978) (tonnes)

Product/region	*United States*	*W. Europe*	*Japan*
Ethylene	2,200,000 (Union Carbide)	1,100,000 (ROW)	770,000 (Mitsui)
LDPE	680,000 (Union Carbide)	840,000 (ROW)	260,000 (Mitsubishi)
HDPE	280,000 (Soltex)	325,000 (Hoechst)	226,000 (Mitsui)
PP	520,000 (Hercules)	320,000 (Montedison)	196,000 (Mitsubishi)
TPA/DMT	1,360,000 (Amoco)	360,000 (Dynamite)	275,000 (Teijin)

Note: Based on data obtained from *Chemical Age*, 1979, various issues.

plastics products accounts for about 50 per cent of the total market, and the largest eight producers for over 80 per cent. It is also true that the production in Western Europe is concentrated in fewer companies than in the United States. About eight or ten companies in Western Europe account for most of the production of petrochemical products (ICI, BP, Shell, BASF, Bayer, Hoechst, Montedison, Solvay, Dow, etc.); in each national market, the leadership is assumed by the largest national company.

Union Carbide, for instance, has a dominant position in ethylene and LDPE production in the United States and ROW (a BASF-Shell joint venture) occupies a similar position in Western Europe. Imperial Chemical Industries and BP are the leading petrochemical producers in the United Kingdom. Hoechst has a leading role in chemical production in Western Europe and has a world-wide HDPE capacity of 700,000 tonnes/year, making it the largest single producer of HDPE in the world. There is, therefore, considerable product specialisation on a company level (Tables 2.11 and 2.12). Table 2.12 shows that, although Japan has a larger aggregate production capacity in petrochemicals than individual European producers, its companies, have smaller single product capacities than the European or American companies, which can be attributed to the diverse activities of the Japanese petrochemical companies.

This product leadership attained by the largest companies gives them virtual control over the markets and allows them to set product prices, with which smaller producers soon fall into line; this practice is well-known in the chemical industry and has been seen recently in petrol price setting in the United Kingdom, where Esso determines the price levels by virtue of its size and market share.

Patent monopolisation in most of the major petrochemical products and plastics materials has been eroded by the development of imitative processes by rival firms or engineering firms, which have no production capacities of their own. However, the introduction of new firms into these markets, especially the smaller ones, has become very limited because the increasing rises in capital and feedstock costs for the new, large-size plants that were coming on-stream, are possible only for the limited number of large firms able to make the huge financial commitments required. The very large size of the new chemical plants with the benefits of economies of scale allowed the larger firms to compete more successfully with smaller producers who found their remaining share of the market diminishing continuously as they were slowly squeezed out of the market.

2.7.2. Trade practices of the petrochemical industry

The oligopolistic market structure and oligopolistic trade practices are revealed in the joint ventures concluded between oil and chemical companies and also in the formation of large consortia, involving the development by the largest companies of huge petrochemical complexes close to the sources of cheap energy in various regions of the world, namely, in Alaska, Australia, Saudi Arabia, Canada and Scotland. The present structure of petrochemical markets is difficult for new producers to penetrate significantly partly because of the integration of production operations and partly because the market is organised on the basis of long-term contracts between producers and users.

A small merchant market exists for styrene in Western Europe, but the way in which the business is conducted is so traditional that it makes the entry of new producers into the existing markets very difficult. Hoechst, for example, is a large user of styrene but has no production capacity of its own; its needs have been traditionally supplied by Gulf who owns a totally merchant plant. In Italy, on the other hand, Shell has no production facility for chlorine, which it obtains from Akzo, which in turn buys its fibre grade ethylene glycol from Shell.[18] BASF has a similar arrangement for polypropylene oxide, which it obtains from ROW (a BASF-BP joint venture). These long-term contracts are normally for three to five years and sometimes may be as long as ten years, where product prices are initially agreed upon but could be modified in the future according to, or by relating them, to spot market naphtha prices or product market prices.

Another trade practice, which limits the prospects of entry into the petrochemical market for new and particularly for small producers, is that producers give bonuses to their traditional consumers who exceed their usual intake of materials by a certain percentage. These bonuses are in the form of discount repayments (2–3 per cent of the value of the total amount purchased) which rise with increased intake, effectively reducing the marginal cost of the extra products bought by the consumer. This bonus system is most widely in usc in West Germany.

Petrochemical producers tend to raise their product prices as soon as oil or naphtha price rises are announced; in 1974 product prices for most petrochemicals rose at higher rates than those of oil.[19] But these producers resist any downward trend in prices, even if oil prices are stagnant or declining in real terms, as has been the truth since 1981/83, when petrochemical products' prices continued to rise despite the relative decline in oil prices over this period.[20]

Finally, international trade in bulk petrochemicals, especially in olefins, is very limited since world production and consumption of these products is concentrated mainly in the three producing regions of the world (United States, Japan and Western Europe), which together account for over 80 per cent of production and more than 70 per cent of consumption of petrochemicals. Most of this trade takes place between these three regions with the EEC countries accounting for most of this trade in terms of EEC inter-trade.

2.8. Concluding remarks

The energy price structure made major chemical companies realise their vulnerable position *vis-à-vis* their feedstock suppliers and their dependence on oil majors for these supplies. They responded to the situation by integrating upstream to secure feedstock supplies for their existing capacities. They also reshaped their operations, especially in Western Europe, by closing down old plants and swapping to others to insure their market leadership in certain products, to minimise competition in the more mature products and to boost prices. The chemical companies were preparing for investment in downstream products in the face of competition in plastics and bulk petrochemicals, where rates of return on investment had diminished from their previous high levels, and also in the expectation of oil majors increased participation in future capacities coming on stream.

The oil majors, in their turn, had the opportunity to become more active in the field of petrochemicals, following the oil price rises, due to their control of oil supplies. Their increased cash flows and profits enabled them to raise the financial resources required for large-scale petrochemical complexes that were coming on stream. The availability of new technologies and processes, developed by engineering firms, and the willingness of many major chemical companies to form joint ventures with them, made these investments possible. The oil majors were also finding it more profitable to move down-stream to produce intermediate and final products instead of simply increasing and expanding their production of basic products such as ethylene and ammonia, since many of these products were already mature and their returns on investment very low.

This period was also marked by the emergence of a new group of government-backed companies, particularly in Italy and France. These companies were able to withstand the new situation since their association with government-controlled oil companies guaranteed their feedstocks.

3 THE BRITISH PETROCHEMICAL INDUSTRY

3.1. Introduction

This chapter discusses how economic theory, namely the *input–output techniques*, could be applied to the study of the petrochemical industry in general and the plastics materials industry in particular. It should be emphasized that the theory serves as a tool for understanding the operations of industry, therefore allowing for better control and organisation of its development process.

The first part of this chapter analyses the structure of the United Kingdom market for plastics materials. Using the powerful input–output (I/O) techniques, the industries that are major users of plastics materials and those that depend most on the use of plastics in meeting their bills of final products, are identified. The prospects of future growth of markets and changes in technology and the implications for the plastics industry and the demands for its products are studied within the framework of input–output models, which allow for technological changes in terms of the use of factor inputs and the market growth prospects to be incorporated into the projected input–output matrix of technical coefficients.

In the second part of this chapter, the supply pattern of the major plastics materials in the United Kingdom and the roles played by the major petrochemical producers are studied. The structural changes that have been taking place on the British scene recently, the performance and the major problems facing the British petrochemical industry are also looked at.

3.2. Estimation of the derived-demand for synthetic resins and plastics materials industry (29)[21] in the United Kingdom: input-output approach

In this section, the extent to which the output of synthetic resins and plastics materials depends on the demands for other industrial sectors as well as the elements of final demand is studied, using the techniques of input–output analysis. The structural patterns of the synthetic resins and plastics materials industry are analysed, in particular the allocation of its output to its major markets. The relative importance of these markets is looked at in two ways. First, those that are direct purchasers of synthetic resins are obtained from the inter-industry transactions Table D[21] then the sectors which are important purchasers of synthetic resins, when direct and indirect purchases are considered, are obtained from the *Leontief inverse matrix*, i.e. Table E[22] of the input–output tables.

This type of analysis is particularly suitable for an industry whose production is used mainly as an intermediate factor input by other industries. It can be developed into a derived-demand relationship between the output of industry (29) and the final demand components. Thus the impact of changes in final demand (e.g. increased exports or consumer expenditure, etc.) on the output of industry (29) could be estimated. This point will be dealt with later on, in the discussion on forecasting demands for industrial outputs for an anticipated pattern of economic conditions.

3.2.1. The allocation of the output of synthetic resins and plastics materials to other industries

The following analysis is based on the most recent input–output tables available for the United Kingdom, which refer to the Census of Production for the year 1974 and the associated purchases inquiry.

Considering the inter-industry flows, Table D[23], the direct sales of synthetic resins and plastics materials (29) to other industries could be traced to find the major purchasing industries of these products and their importance as inputs to the major industries; thus determining the industries that most depend on the use of plastics materials to meet their bills of final outputs.

The development and the growth of these industries would determine the demands for the products of industry (29), which is mainly a derived-demand since very little of these products is consumed directly by final demands as their consumption is accounted for by other industries as intermediate products.

In 1974, the total output of synthetic resins and plastics materials had a value of over £976 million; two-thirds of this output was used by other industries as intermediate input and the remaining one-third went to final demand, mainly for export (over 90 per cent of final demand) for further processing (see columns 1 and 2 of Table 3.1).

Table 3.1 lists twenty-five of the major purchasers of the output of sector (29). The first column of the table lists the direct sales of plastics materials in millions of pounds, which range from £162.8 million purchased by the largest direct user, the plastics products industry (90), to £0.3 million going to the motor vehicles industry (59). Ten sectors accounted for purchases of over £20 million each, namely paint (27), other chemicals (32), insulated wires and cables (49), man-made fibres (68), paper packaging products (86), rubber (89), plastics products (90), other manufacturing (91) and construction (92). The synthetic resins and plastics materials industry (29) itself was the third largest buyer with an intra-industry purchase of £44.8 million.

A clearer picture results if the market share of these industries is considered as a percentage of the total output of industry (29) instead of absolute figures of direct purchases. These are listed in column 2 of Table 3.1. It is apparent that a few of these industries account for a considerable share of the market. Plastics products (90) alone accounts for over 16 per cent of the total market, putting it in the first place as a major consumer of plastics materials and synthetic resins, followed by rubber (89), synthetic resins and plastics materials (29), construction (92) and paint (27), etc. as shown in column 3 of Table 3.1.

The twenty-five industries listed in Table 3.1 account for about 50 per cent of the total output of industry (29) in direct purchase, or 87 per cent of the total intermediate output (104).

3.2.2. Synthetic resins and plastics materials (29) as input coefficients to twenty-five industries

Input coefficients indicate the importance of the components of factors of production to the final product of each industry. Sometimes these are referred to as technical coefficients since they express the combination of the inputs of each industry according to the technology of the predominant process in that industry.

The first column in Table 3.2 shows the amount of synthetic resins going into each industrial sector as a percentage of the intermediate inputs of these sectors. The industries that are intensive direct users of synthetic resins and plastics materials, i.e.

Table 3.1 Allocation of synthetic resins and plastics materials output to twenty-five major purchasing industries and final demand, for the United Kingdom, 1974

		Output distribution	*Output coefficient i.e. in direct purchase*		*Direct & indirect purchase of plastics materials by final demand**	
SIC No.	*Sector*	*(£ m)*	*%*	*rank*	*(£ m)*	*rank*
13	Bread etc	8.6	0.88	16	12.4	14
24	General chemicals	10.9	1.11	13	7.8	. .
27	Paint	31.7	3.25	5	12.23	15
29	Synthetic resins	(44.8)	4.40	3	362.48	1
32	Other chemical industries	23.5	2.40	8	15.86	8
47	Instrument engineering	6.0	0.61	20	11.65	16
49	Insulated wires and cables	23.0	2.35	9	5.14	. .
51	Electrical components	10.3	1.05	14	7.51	. .
52	TV and radio equipment	7.6	0.78	17	11.55	17
55	Domestic electrical goods	9.4	0.96	15	13.17	12
56	Other electrical goods	6.0	0.61	21	3.68	. .
59	Motor vehicles	0.3	0.03	. .	23.27	5
68	Man-made fibres	25.0	2.56	6	13.50	10
72	Carpets	12.5	1.28	12	12.62	13
75	Other textiles	14.9	1.52	11	5.24	. .
78	Footwear	6.4	0.65	19	11.37	18
84	Timber and wood manufactures	5.6	0.57	23	3.35	. .
85	Paper and board	6.6	0.67	22	2.15	. .
86	Paper packaging products	24.8	2.54	7	1.41	. .
87	Other paper products	7.0	0.72	18	3.41	. .
89	Rubber	56.4	5.78	2	21.28	7
90	Plastics products	162.8	16.67	1	30.40	4
91	Other manufactures	24.9	2.55	. .	21.93	. .
92	Construction	38.4	3.93	4	97.21	2
98	Distributive trade	13.5	1.38	. .	40.64	3
104	Total intermediate	615.6	63.05	—	—	—
105	Consumer's expenditure	3.2	0.33			
106	Government consumption	1.6	0.16			
107	Fixed capital formation	5.5	0.56			
108	Stocks	24.7	2.53			
109	Exports	325.9	33.37			
110	Total final output	361.0	36.95			
111	Total output	976.6	100.00			

* Obtained from Table C1 (Appendix C).

. . Insignificant.

Table 3.2 The importance of the output of synthetic resins and plastics materials industry (29) in relation to the structure of inputs of twenty-five major industries, United Kingdom, 1974

		Input coefficients for industry's (29) purchasers			*Inverse coefficients for industry's (29) purchasers*	
SIC No.	*Industry*	*Intermediate input %* *(1)*	*Total input %* *(2)*	*Rank* *(3)*	*%* *(4)*	*Rank* *(5)*
13	Bread etc.	1.43	0.77	18	1.20	19
24	General chemicals	1.07	0.48	21	0.69	24
27	Paint	14.96	8.70	2	9.11	3
29	Synthetic resins	8.12	(4.59)	4	(100.41)	1
32	Other chemical industries	6.68	3.36	9	3.97	8
47	Instrument engineering	1.75	0.70	20	1.53	18
49	Insulated wires & cables	6.38	3.43	8	3.80	9
51	Electrical components	3.63	1.53	14	2.16	14
52	TV and radio components	3.07	1.37	15	2.15	15
55	Domestic electrical goods	3.43	1.98	12	3.21	11
56	Other electrical goods	2.10	0.91	17	1.59	17
59	Motor vehicles	0.02	0.008	25	0.80	22
68	Man-made fibres	8.13	4.22	5	4.64	5
72	Carpets	5.13	2.87	10	3.78	10
75	Other textiles	7.69	3.45	7	4.08	7
78	Footwear	3.69	1.64	13	3.19	12
84	Timber and wood manufacture	1.65	0.48	21	0.78	23
85	Paper and board	1.98	0.73	19	1.09	20
86	Paper packaged products	5.92	2.62	. .	3.15	13
87	Other paper products	2.84	1.13	16	1.64	16
89	Rubber	14.93	6.36	3	6.70	4
90	Plastic products	36.64	17.43	1	17.84	2
91	Other manufactures	6.94	3.66	6	4.64	5
92	Construction	0.93	0.35	23	1.01	21
98	Distributive trade	0.31	0.10	24	0.43	25
104	Total intermediate	—	—	—	—	—

( ) Estimates.
. . Insignificant.

* Direct and indirect synthetic resins and plastics materials inputs per £100 of final demand for the output of the purchasing industries.

Note: Columns 1 and 2 obtained from Table D, *Input–Output Tables for the UK, 1974*, Business Monitor PA 1004, 1981. Column 4 obtained from Table E, *Input–Output Tables for the UK, 1974*, Business Monitor PA 1004, 1981.

where intermediate inputs are concerned, are clustered around three industrial sectors – the chemical industry itself: paint (27), synthetic resins and plastics materials (29), other chemicals (32), rubber (89) and plastics products (90); the textiles and related industries: man-made fibres (68), other textiles (75) and carpets (72); and, finally, the electrical industry: electric components (51) and insulated wires and cables (49). These are the sectors that are most dependent on the use of the products from sector (29) in meeting their bill of final outputs.

On the other hand, column 2 of Table 3.2 gives the direct use of polymers as a percentage of the total output of each industry. Thus relating the direct demand for synthetic resins required by each industry to the final output of that industry. By determining the important industrial users of plastics materials, the changes in the outputs of these users resulting from the growth of their markets can then be used as an indicator of the demands for the products of sector (29).

The intensiveness in the use of plastics materials by major users is best depicted by ranking these industries, shown in column 3 of Table 3.2, according to the results of column 2. The input of plastics materials and synthetic resins into the plastics products industry (90) constitutes more than 17 per cent, at the top of the list as the industry most dependent on the supplies of sector (29), followed by paint (27), rubber (89), synthetic resins and plastics materials (29), man-made fibres (68), other manufacturing (91), other textiles (75), insulated wires and cables (49), other chemical industries (32), carpets (72) and paper packaging and products (86). With the exception of the plastics products industry (90), synthetic resins and plastics materials inputs account for between £2.5 to £8.7 per output of £100 each of the above industries (see column 2, Table 3.2).

3.2.3. Inverse coefficients for synthetic resins and plastics materials' purchasers

Table E of the *Input-Output-Tables* [24] is the table of the inverse coefficients of the *Leontief matrix*, which represent the solution of the equations system for the open static input–output model (Leontief, 1951).

The inverse coefficients for the major purchasers of synthetic resin and plastics materials (29), listed in column 4 of Table 3.2, are obtained from row 29 of Table E. They show the direct and indirect plastics materials inputs to other industries in coefficient form, so that linkage effects are taken into account; that is, inverse coefficients express the direct inputs going into the production of final products of the industries concerned and the indirect inputs which are the components of intermediate products that are themselves users of the products of industry (29). A good example illustrating this process of linkage effects is the motor car industry which uses components supplied by other industries, such as paint, wires and cables, instrument panels and many other parts, which include a significant amount of synthetic resins and plastics materials. The indirect inputs in this case account for a much higher amount of synthetic resins than those consumed directly by the motor car industry.

Comparing the inverse coefficients in column 4 with the direct input coefficients of column 2 in Table 3.2 shows that most coefficients increase slightly when indirect effects are considered. In some cases the increase was substantial in relative terms, amounting to a threefold increase for construction (92), rising from £0.35 (column 2) of direct plastics inputs per £100 of total output of construction to £1.01 (column 4). Similarly for the distributive trade (98), the increase was from 0.10 to 0.43 per cent, for footwear (78) from 1.64 to 3.19 per cent and for domestic electrical goods (55) from 1.98 to 3.21 per cent. While these increases were substantial in relative terms, that is, compared to initial values, their absolute values were still small (less than 5 per cent) in proportion to the total input requirement of these industries, which left their overall ranking relatively unchanged (compare columns 3 and 5 of Table 3.2), with the exception of the synthetic resins industry (29) itself and the motor vehicles industry (59), both of which show a striking difference from the other industries.

Industry (29) has an inverse coefficient of 100.41, putting it in the first rank as a direct and indirect user of its own products. The inverse coefficient of more than 100 per cent implies that every £100 of final demand for products of the industry induces an overall output of £100.41 of synthetic resins and plastics materials, which can be broken down into: £100 of final demand for products of synthetic resins and plastics materials and £0.41 of intra-sectoral purchase of synthetic products within the industry itself, required by the industry to produce a final output of £100.

The motor vehicles industry (59), on the other hand, has a direct input of £0.008 of synthetic resins and plastics materials per £100 of total output, which increases by 100 times to £0.80 for every £100 in final demand for motor vehicles when the direct and indirect inputs are considered. However, this value of the inverse coefficient is still very small relative to other industries and it improves the ranking of the motor vehicles industry (59) only slightly, from 25 (column 3) to 22 (column 5, Table 3.2).

The significance of these calculations is highlighted when the effects of final demand components on the demand for plastics materials are taken into consideration, as explained below.

3.3. The dependence of the production of synthetic resins and plastics materials industry (29) on other industries according to their final demand components

The intermediate demand for plastics materials can, as seen above, be best obtained by taking into consideration the direct and indirect demand for plastics materials by other industries. It was also seen that the indirect demand could be more important than the direct demand as in the case of the motor vehicles industry (59), construction (92) and the distributive trade (98).

This section and the following section analyse the importance of final demand in stimulating industrial production (29) directly and indirectly and the varying importance of the major industrial users of plastics materials is assessed according to their main final demand components. This is done by multiplying the row of the inverse Leontief matrix corresponding to industry (29) by the matrix of the final demand components of all industries, as shown in matrix form below:

$$X = (I - A)^{-1} F$$

where X = the matrix corresponding to the total output of each industrial sector;

$(I-A)^{-1}$ = the Leontief inverse matrix, which corresponds to Table E of the input–output tables;[21]

F = the matrix of final demand components.

Table 3.3 Gross demand for the output of synthetic resins and plastics materials industry (29) by the ten largest buyers in accordance with their major final demand component for the United Kingdom, 1974

		Final demand by major component column			*Total final demand*
Rank	*Industry (SIC)*	*Sector*	*£ m.*	*%*	*£ m.*
1	Synthetic resins and plastics materials (29)	exports	327.24	90	362.48
2	Construction (92)	gross fixed capit. form.	76.00	78	97.21
3	Distributive trade (98)	consumers' expenditure	33.63	82	40.64
4	Plastics products (90)	consumers' expenditure	14.83	49	30.40
5	Motor vehicles (59)	exports	11.03	47	23.27
6	Other manufacturing (91)	exports	10.20	46	21.93
7	Rubber (89)	exports	13.80	65	21.28
8	Other chemical industries (32)	exports	9.71	61	15.86
9	Other services (102)		6.98	48	14.41
10	Man-made fibres (68)		12.29	91	13.50
	Total (1 to 10)				640.98
Gross output of synthetic resins and plastics materials industry (29)					976.00

Note: Obtained from Table C1 (Appendix C).

The detailed results of this exercise are given in Table C1 (Appendix C) and summarised in column 4 of Table 3.1. Also, the ten major industrial users of plastics materials and the main inducing final demand components are given in Table 3.3.

The fourth column of Table 3.1 lists the major markets for synthetic resins and plastics materials, which are arranged according to the total (direct and indirect) expenditure on plastics materials induced by the final demand for the products of the industries concerned. The largest five of these, which have total inputs of plastics materials valued at over £20 million each, were industry (29) itself, construction (92), distributive trade (98), plastics products (90) and motor vehicles (59).

Synthetic resins and plastics materials (29) and plastics products

(90) respectively, the first and fourth major markets by final demand as indicated above, are both important intensive-direct users of plastics materials (see their ranking in column 3 of Table 3.1), which makes them dependent on sector (29) in meeting their output requirements.

Construction (92) and distributive trade (98), respectively the second and third most important markets for plastics materials by final demand, show a different consumption pattern in using plastics materials from the previous two sectors. While construction and distributive trade are among the first ten major direct markets for plastics materials (column 3, Table 3.1), they are both non-intensive users of plastics materials, as shown from their ranking in columns 3 and 5 of Table 3.2, putting them at the bottom of the scale, whether direct or total (direct and indirect) input requirements are considered. However, this pattern changes completely when the importance by final demand is looked at, with construction (92) assuming the second major market and distributive trade (98) the third. This is due to the sheer size of these two industries and their importance to final demand expenditures.

On the other hand, the motor vehicles industry (59) has a somewhat different pattern of consumption of synthetic resins from the first two types of industries. Industry (59) is one of the least industries in purchasing directly from sector (29) with purchases amounting to 0.03 per cent of the total output of sector (29) (see column 2, Table 3.1). Industry (59) has an inverse coefficient of 0.8 per cent which still puts the industry at the bottom of the scale of input coefficients with a rank of 25 (column 5, Table 3.2).

However, the rank of motor vehicles (59) as a market for synthetic resins and plastics materials (29) jumps to the fifth place, making it one of the major markets by final demand. The anomalous position of this industry is due to the derived-demand pattern for plastics materials through the use in cars of various components which are intensive users of plastics materials, as explained earlier. The significance of this industry as a market for plastics materials and its future potential has been discussed in more detail in Chapter 2.

3.4. Demand for synthetic resins and plastics materials by final demand

The consumption of synthetic resins and plastics materials (29) by the various sectors of the economy, induced by final demand, is given in Table C1. The most important factor of final demand in influencing the demand for the output of industry (29) is that of export which accounts for about £512.3 million or over 50 per cent of the total output. This includes direct exports of synthetic resins and plastics materials in the intermediate form (£327 million) and the rest indirectly via the export of finished goods that include plastics materials in their manufacture.

The second most important factor is that of consumers' expenditure amounting to £257 million (last row in Table C1), or about 25 per cent of the total output of synthetic resins and plastics materials.

The ten largest markets by final demand for plastics materials are arranged in Table 3.3, which also indicates the major final demand components for each. These ten industries account for about two-thirds of the total output of industry (29), with exports as the predominant factor of final demand for six of the ten industries. These are synthetic resins and plastics materials (29), motor vehicles (59), other manufacturing (91), rubber (89), other chemicals (32) and man-made fibres (68).

The construction industry (92), being the second largest market by final demand, was largely dominated by the demand for gross fixed capital investment.

Finally, the demand by the motor vehicles industry was mainly for exports but also, almost equally, for gross fixed capital investment. Whereas for distributive trades (98) and plastics products (90) the demands went mainly for consumers' expenditure.

3.5. A methodology for assessing future demands for petrochemical products

Petrochemical products are almost entirely produced to serve as intermediate products to other industries; therefore, demand for these products is best estimated as a derived demand through the prospects of growth of the major purchasing industries, which will determine the future demands for petrochemical products.

The intersectoral analysis of the previous sections (3.2–3.4) defines this derived-demand relationship which exists between the output of a petrochemical product and its end-use markets and the final demand components. It also shows how the importance of each major purchasing industry could be determined, in terms of the volume of purchase of plastics materials, their intensiveness of use and its relation to the total output of the purchasing industry.

The domestic industrial output of petrochemicals depends on the economic environment, resulting from the growth prospects of the Gross Domestic Product in general, government policy and public expenditure for gross fixed capital investment, the level of import penetration of manufactured goods and consumers' expenditure. Exports are affected by exogenous factors such as world trade, but also equally by the exchange rate and the level of taxation of energy resources and their overall costs, which are influenced by government policy. It is, therefore, more appropriate to relate the output of the petrochemical industry to the economy as a whole, where the interindustry flows can be more readily obtainable from the separate consideration of each of its major markets via input–output techniques instead of simply relating its output to Gross Domestic Product or even the aggregate changes in each of the major components of Gross Domestic Product.

The importance and usefulness of the input–output models is that they can be linked to a macro-economic model which projects the overall economic outset for the period under investigation.

Woodward (1980) and the United Kingdom Department of Industry developed a seventy-sector input–output model for monitoring the forecasting developments in the manufacturing industry. The model consists of two parts: the first includes the results of the macro-economic model which reflects the economic environment resulting from government policy, the state of the economy and its sectoral relationship, and the world economic outlook. The second part of the model includes the input–output and 'make-matrix' tables. The model ensures consistency between the demand and supplies of the industrial products under study, subject to the conditions determined by the macro-economic model, namely consumers' and government expenditure and fixed investment on the domestic level. On the other hand, world trade, prices of imports and their level of penetration are the main exogenous economic factors taken into account by the model which is fully computerised to allow rapid evaluation of alternative scenarios.

A similar model could therefore be developed for use as a forecasting mechanism to project outputs, growth rates of production and prospects for individual petrochemical products, provided the input–output table can be successfully disaggregated to a detailed commodity level for the products concerned and a proper updating procedure can be carried out on the input–output table, since the proposed model is dynamic in nature.

Disaggregation to a detailed commodity level within the input–output table needs to be carried out only for those industries associated with petrochemical products and their major users so that the linkage effects between the individual products and their end-use markets could be related to the components of final demand.

For instance, increased (or cuts in) government spending on housing, health and educational facilities would have an enormous effect on the construction industry as well as investment in transportation services, communications and defence increases, the demand for motor vehicles, electrical components and their related industries. Consumers' expenditure affects two important industries, motor vehicles and packaging. All these industries are major markets for plastics materials, as has been seen in Sections 3.2.3 and 3.3.

The disaggregation of final demand components into expenditures on industrial sectors or their final products (which is in effect the estimation of the derived-demand equations for the petrochemical products under study) is based on past trends and regressions of time series data, together with any special allocations or developments expected for particular industries. A more detailed treatment of this subject area can be found in the writings of Liebeskind (1972), Stone and Woodward (1980), who carried out extensive research on projecting developments in the chemical and electrical industry using input–output disaggregation techniques within the framework of input–output models.

3.5.1. Projecting input–output tables

The input–output matrix of coefficients can be updated every year to allow for technical change or increased substitution between inputs in the process of production for each product or industry. As mentioned earlier, the projected input–output matrix can be introduced into the model, making it a dynamic model in that it relates future developments to what happened the year before.

The projection of input–output matrices was studied extensively by the Cambridge Growth Group.[25] A variety of methods were used to account for the changes in input–output coefficients over time, resulting from technological change, a change in the mix of products of a non-homogeneous output or due to economies of scale in the use of some inputs, (where marginal input coefficients fall as economies of scale are increasingly exploited with the growth of output levels). Another important factor is the change in relative prices of inputs over time. These methods can be grouped into four categories.

1. trends in coefficients;
2. updating the base matrix;
3. restricted price substitution; and
4. non-homogeneous production functions.

T.S. Barker (1977b) tested the above methods for the United Kingdom input–output tables; his findings confirm the need to use the most recent input–output table it is possible to construct, using an updating procedure, developed by the Cambridge Group, while keeping in mind that certain groups of coefficients behave in a characteristic manner, and incorporating industrial expertise wherever possible, for any known or expected changes in the updated coefficients. Using two independently obtained input–output tables, Barker found that the projected coefficients had improved, especially when own-price elasticities of intermediate products were included, according to their characteristic group behaviour.

3.5.2. Input–output models versus regression models

It is generally accepted that input–output models give better projections than regression models in the short run, while there is evidence that regression models are better at projecting industrial outputs in the long term (Barnett, 1966). A study by Ghosh (1966) on United Kingdom industry for the years 1949–55 confirms that input–output models are better at projecting industrial output, particularly for industries where final demand is only a small proportion of total output, than other models such as regression, final demand blown-up or Gross National Product blown-up models. This is because regression models use a small number of variables that correlate with the projection, whereas input–output models allow a very large number of variables to influence the projection but they do that in a simple and crude way which gives rise to the accumulation of errors as the period of projection lengthens. The major disadvantage of input–output models is that they do not take into consideration price effects of complementary materials and substitutability of inputs. However, the accumulation of errors in input–output models, as the period of forecast becomes longer, can be overcome by continuously adjusting and updating the matrix of coefficients.

So, in addition to industry-by-industry effects on the output of the sector under consideration, being allowed for by input–output models, the model itself can be modified to take account of any changes that have taken place or are expected to do so in future. These changes and the inter-industry effects are more visible and are easier to investigate with input–output than with other types of models due to the structure of the former, where changes in outputs and demands are worked throughout the *system* as inter-industry transactions.

3.6. The supply of petrochemical products and plastics materials and assessment of the British situation

Having dealt with the demand side of synthetic resins and plastics materials, using input–output techniques, in the previous sections of this chapter, we turn here to major products, the development of production capacities and ownership by major chemical producers, the British position relative to its European counterparts and the recent developments in the process of restructuring the industry on the British scene. Also, developments in two major purchasing industries, construction and the car industry, and their impact on the production growth of the plastics industry are studied.

3.6.1. The production of plastics materials and petrochemical products in the United Kingdom

The development of the petrochemical industry in the United Kingdom followed a pattern similar to that of the other major chemical producers of the industrialised countries, where synthetic materials were supplying and continuously substituting naturally occurring materials whose production has its known constraints and consequent fluctuations in prices. These synthetic materials were also making inroads into conventional markets as well as new applications and new markets that were growing rapidly during the 1950s and 1960s in the construction boom period after the Second World War, where annual industrial growth averaged over 6 per cent in the industrialised countries. The presence of a well-established chemical industry with a sound R & D base (ICI) and the access to cheap oil supplies from the Middle East (BP) allowed the rapid expansion of the petrochemical output to supply the growing demand of the consuming industries and the export markets.

Since the early sixties, the rate of growth of the British chemical industry has consistently been more than double that of the manufacturing average and of the Gross Domestic Product. Return on capital has been high and capital investment for petrochemicals has been higher than those of many other industries. In 1975 the chemical industry contributed 9 per cent of the total Gross Domestic Product, making it the fifth largest industry, yet it only accounted for 5.2 per cent of all employment, making it the eleventh largest employer (NEDO, 1981). The production of organic chemicals, polymers and plastics materials, were the fastest growing sectors, which together accounted for most of the petrochemical output or about 35 per cent of all chemicals.

From the 1950s until the early 1970s, the production of petrochemical products was increasing at a rapid rate of 15–20 per cent each year for many products, with production almost doubling every five years. The output of plastics materials by the mid-1970s was ten times what it had been twenty years earlier. By 1970 the average annual growth rate of plastics was about 10 per cent (1963–73) but many products were growing at higher rates, in particular polyolefins.

Four plastic products, PE, PVC, PS and amino plastics, account for most of the production of synthetic materials. They show a

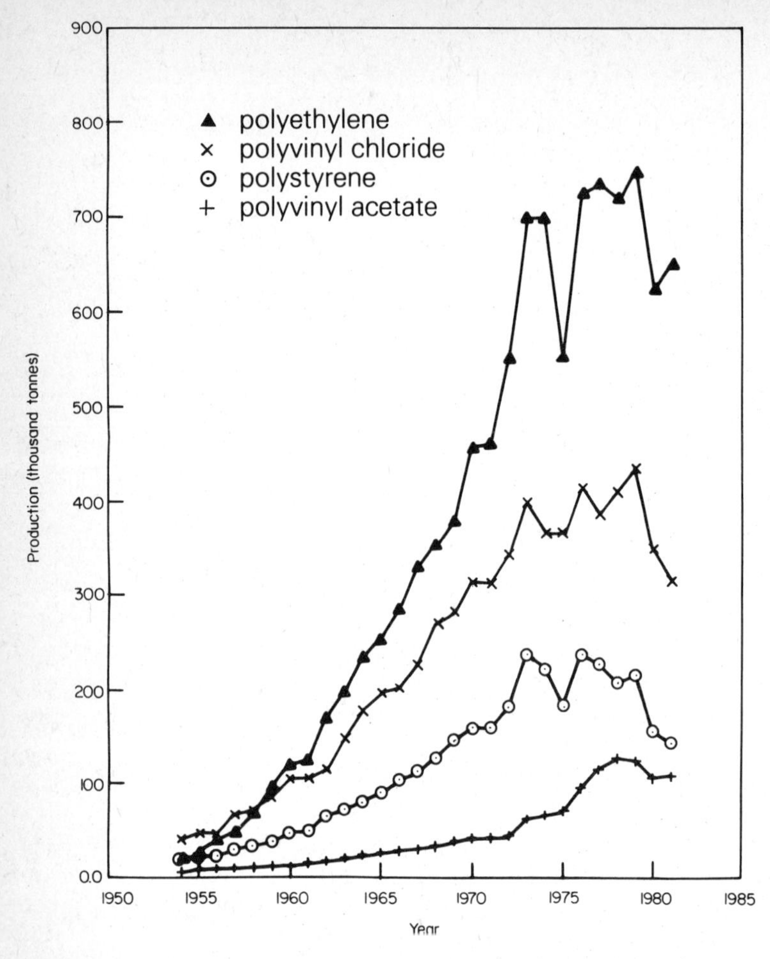

Figure 3.1 Production of the major plastics materials in the United Kingdom

Note: Data obtained from Production series P21, *Synthetic Resins and Plastics Materials*, Business Monitor PQ2 76.1, (London: HMSO, various issues).

pattern of rapid increase in production growth at the early stages of their introduction into the market; their outputs, thereafter, continue to increase but at lower growth rates than before, except for the products with wide ranges of application or those supplying substantial proportions of their outputs to major industries with

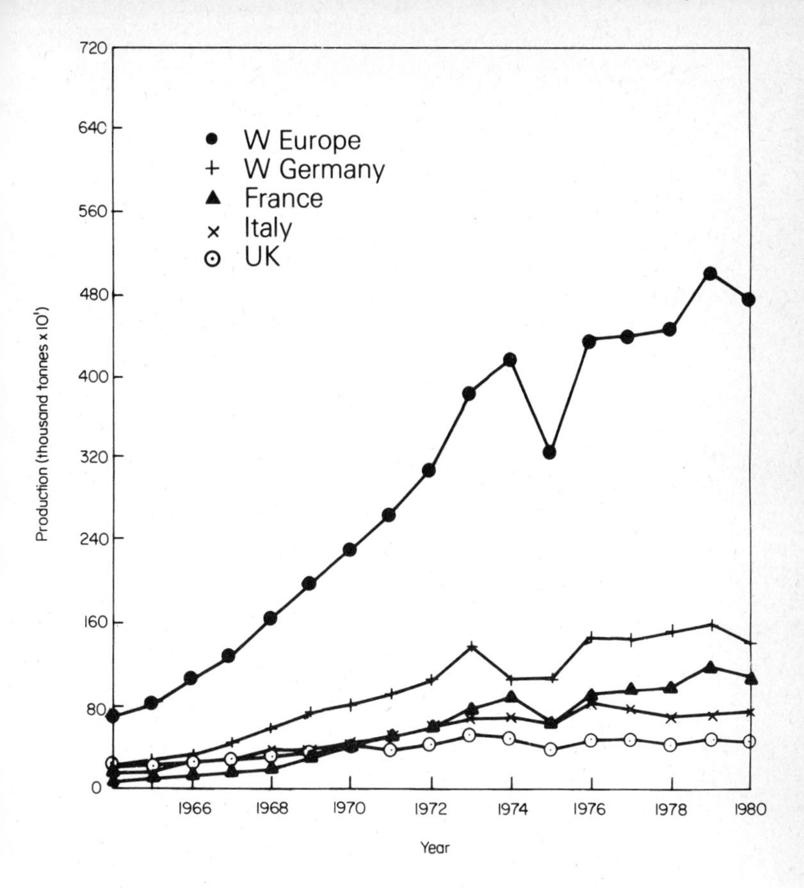

Figure 3.2 Polyethylene production by main Western European producing countries and total Western Europe

Note: Based on data obtained from *United Nations: Yearbook of industrial statistics*, various issues.

good prospects of growth. The competitive prices, good performance and superior physical properties of these materials ensured the continuous growth of their outputs.

Figure 3.1 shows the growth of outputs of the major plastics materials in the United Kingdom since 1954. The relative share of the total market for each product is also apparent from the linear plot. The production curves of these plastics materials are similar in

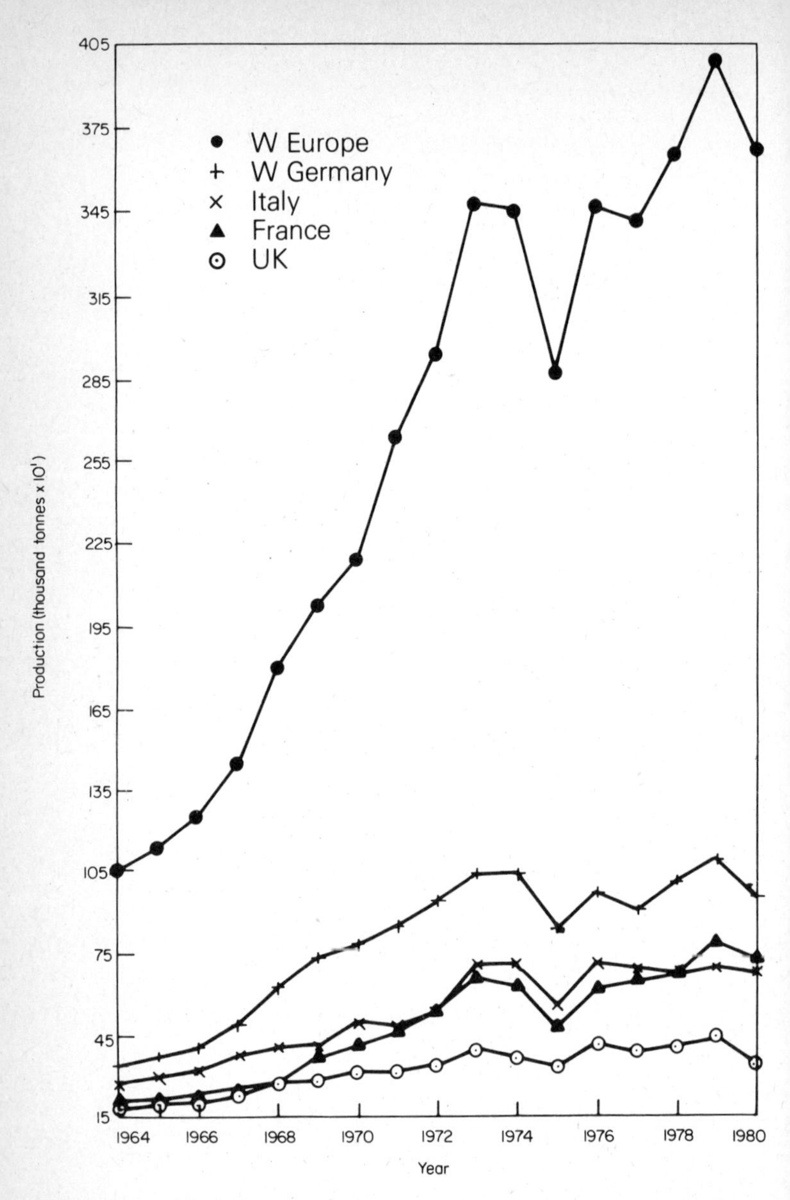

Figure 3.3 Polyvinyl chloride production by the main Western European producing countries and total Western Europe

Source: See note in Figure 3.2.

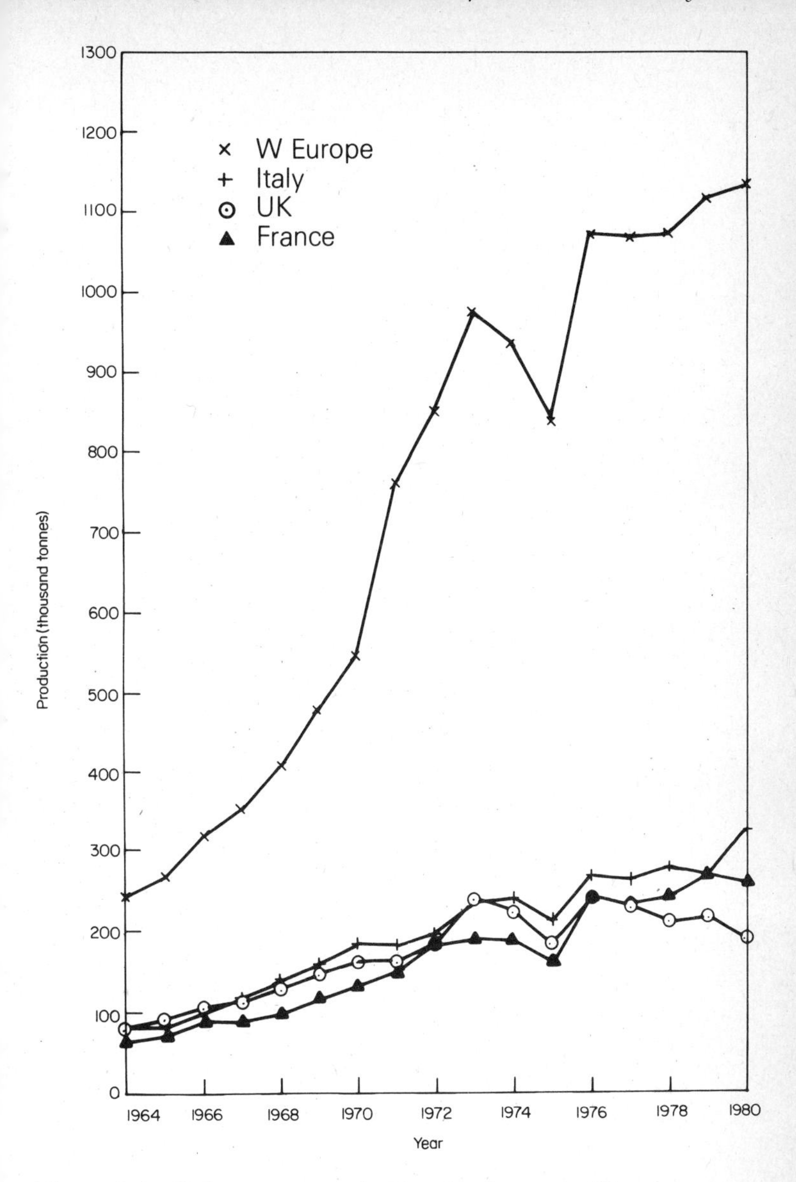

Figure 3.4 Polystyrene production by the main Western European producing countries and total Western Europe

Source: See note in Figure 3.2.

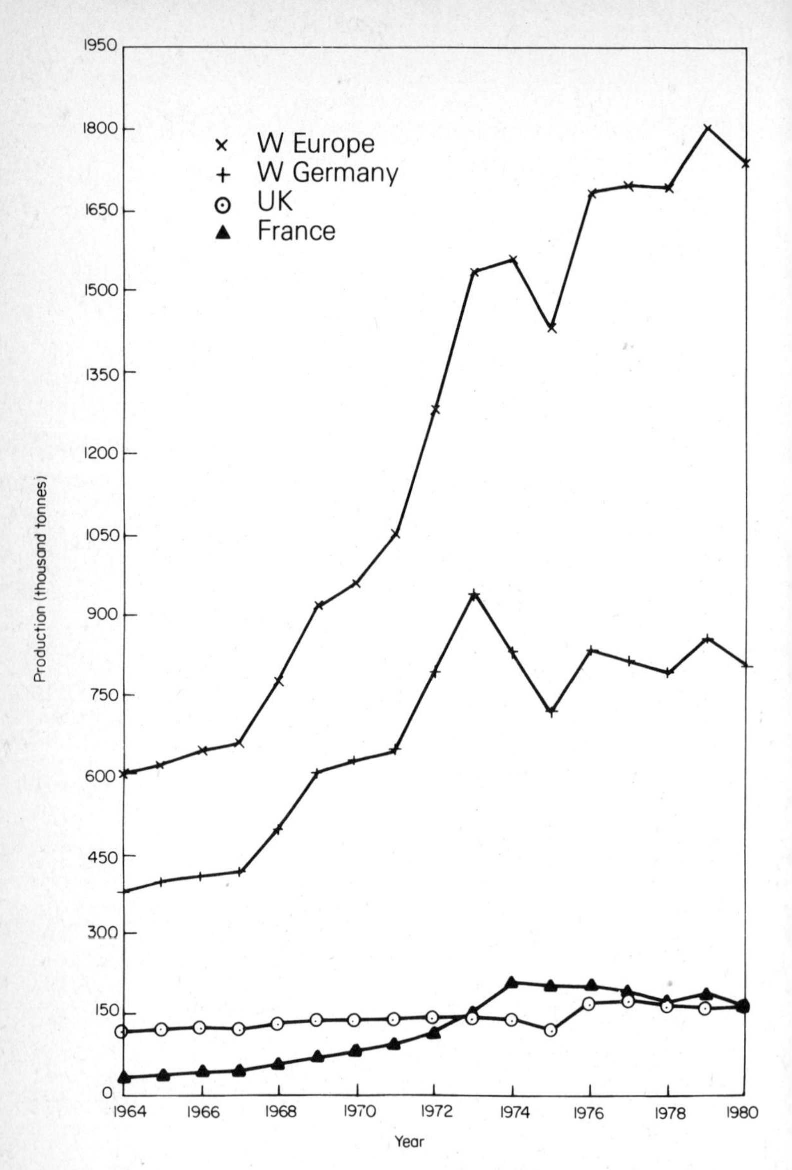

Figure 3.5 Amino plastics production by the main Western European producing countries and total Western Europe

Source: See note in Figure 3.2.

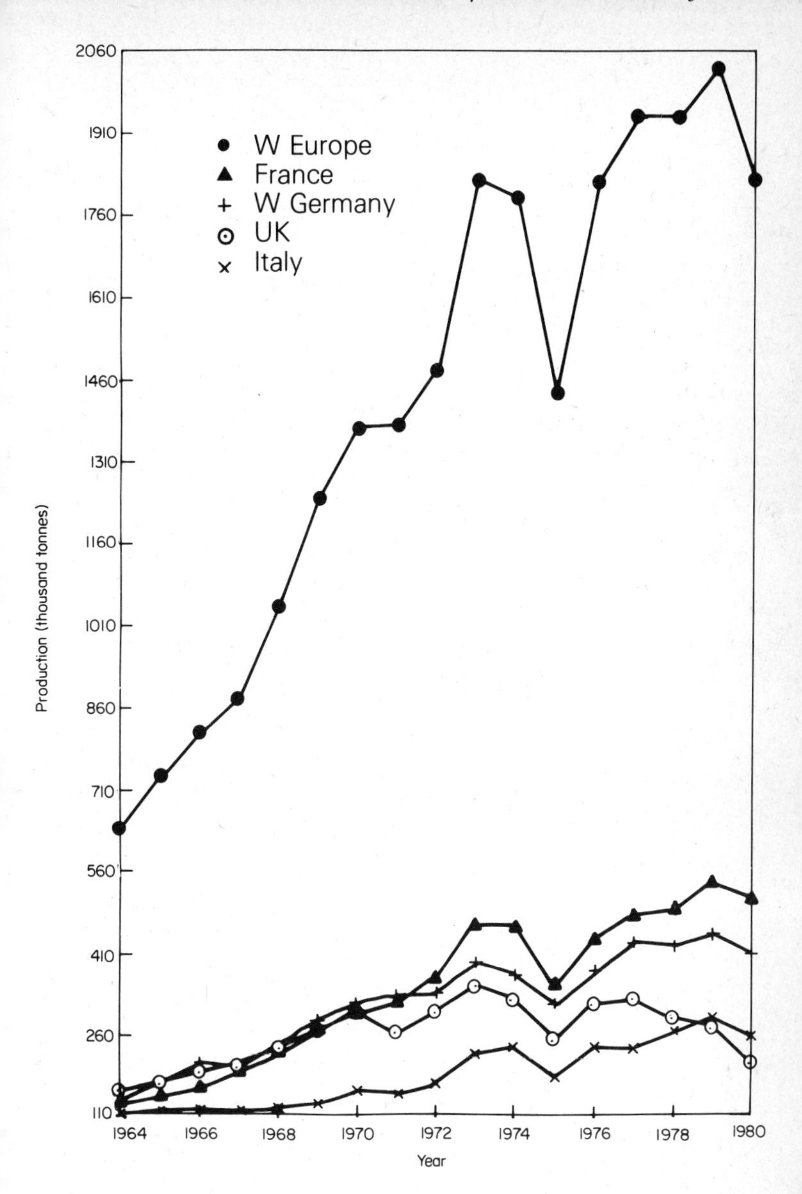

Figure 3.6 Synthetic rubber production by the main Western European producing countries and total Western Europe

Source: See note in Figure 3.2.

shape, thus showing a similar historical pattern of growth, where until 1973 the average annual growth rate for each product was over 12 per cent, with the largest volume products, PE and PVC, growing at rates of over 15 per cent each. However, after 1973 the growth of these materials slowed down relative to the previous experiences of the past two decades; growth on average was just over 2 per cent for PE, PVC and PVAC, but was negative for PS. After 1979, production was actually declining for all products, back to the levels of the early 1970s.

Figures 3.2 to 3.6 clearly show that, while the British industry enjoyed a healthy growth until the early 1970s, other European producers were growing at higher rates than those of the United Kingdom. In real terms, the United Kingdom's share of the total Western European production over the period 1964–80 for all major synthetic products has declined to half its previous level, from around 20 per cent (30 per cent for PE) to less than 10 to 16 per cent.

After 1976, the production in Germany and France, with the exception of amino plastic in France, was growing steadily for all products. Italy's position was similar to that of the United Kingdom, where growth was sluggish for its products, with the exception of polystyrene which continued to grow at a steady rate.

However, since 1973, the British production of polystyrene, amino plastics and synthetic rubber, was showing a trend of slow decline which continued into 1980 (see Figures 3.4, 3.5 and 3.6).

The British share of HDPE and PP production were 4 per cent and 30 per cent respectively in 1975. These shares were later on adjusting to market growth, and as more plants came on stream (in Europe and the United Kingdom) for these two fairly new products, the British HDPE share rose to 7 per cent while that of PP dropped to 18 per cent of the total Western European production by 1980.[26]

Overall, the British position as a major producer of plastics materials has weakened over the last two decades. Its share of basic petrochemical products (ethylene, benzene, etc.) has also been declining (see Figures 3.7 to 3.9), but ethylene capacity recovered in 1981, when it rose to almost double the level of 1980.[27] This decline in output and production capacity is also reflected in the deteriorating position of the British trade in petrochemicals and plastics materials with its European partners and the rest of the world.

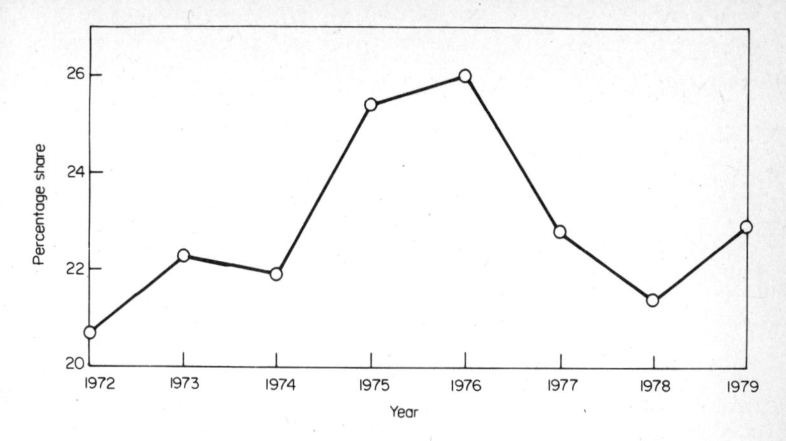

Figure 3.7 British share of Western European ethylene production

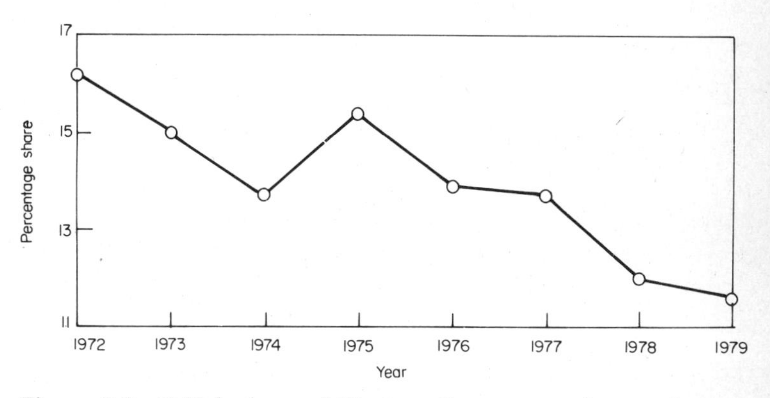

Figure 3.8 British share of Western European polypropylene production

The total Western European production of plastics materials showed a growth pattern similar to that of the United Kingdom and other major producers, where production was increasing rapidly until the early 1970s, which slowed down after 1973 but was still growing at rates higher than the average rates of the traditional producers as a result of the new production capacities that were coming on stream in Greece, Spain, Portugal and the Scandinavian countries (see Figures 3.2, 3.4 and 3.5).

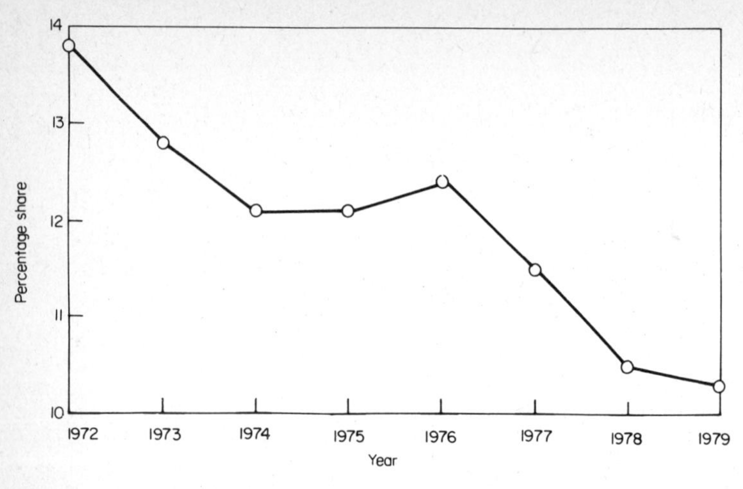

Figure 3.9 British share of Western European benzene production

Source: NEDO, petrochemical report, 1981.

3.6.2. Ownership of the major plastics materials' production capacities in the United Kingdom

The structure of the industry during the 1960s was fairly stable as a result of the nature of the industry with its high capital expenditure, requirements of qualified staff for R & D and a sophisticated marketing meant that the market was captured by a small number of large companies.

In 1969, six companies owned over 90 per cent of the British thermoplastics production capacity, the majority of these companies were organic chemical producers but some were subsidiaries of major oil companies. However, after 1973, the structure of the industry was changing, with oil companies increasing their share of production capacity and the corresponding share of the market.

The ownership of major plastics materials production capacities by producer companies over the period 1969 to 1980 is shown in Table 3.4. The production of thermoplastics in the United Kingdom in 1969 was dominated by ICI with about 40 per cent of the total capacity, and its position was even stronger in PVC, PE and PP; BP was the second largest producer, followed by Shell. The American

Table 3.4 Ownership of main thermoplastic capacity in the United Kingdom, 1969 (000's tonnes/year) and capacity in 1980

Company/prod.	*PVC*	*(%)*	*LDPE*	*(%)*	*HDPE*	*(%)*	*PP*	*(%)*	*PS*	*(%)*	*Total*
ICI	180	(50)	140	(43)	—	—	60	(80)	—	—	380
BP	140	(39)	—	—	44	(5)	—	—	18	(10)	202
Shell	—	—	60	(18)	30	(41)	15	(20)	45	(24)	150
Monsanto	—	—	50	(15)	—	—	—	—	40	(21)	90
BXL	15	(4)	80	(24)	—	—	—	—	—	—	95
Dow	—	—	—	—	—	—	—	—	50	(26)	50
Sterling	—	—	—	—	—	—	—	—	35	(19)	35
Vinatex	25	(7)	—	—	—	—	—	—	—	—	25
Total	360	(100)	330	(100)	74	(100)	75	(100)	188	(100)	1,027
Capacities in 1980 (000's tonnes/year)											
ICI	280	(40)	240	(38)	—	—	190	(56)	—	—	710
BP	230	(33)	—	—	166	(100)	—	—	40	(16)	436
Shell	—	—	160	(25)	—	—	150	(44)	80	(31)	390
Monsanto	—	—	50	(8)	—	—	—	—	(not disclosed)		50
BXL	—	—	187	(29)	—	—	—	—	—	—	187
Dow	—	—	—	—	—	—	—	—	73	(29)	73
Sterling	—	—	—	—	—	—	—	—	60	(24)	60
Vinatex	100	(14)	—	—	—	—	—	—	—	—	100
Brit. Ind. Plastic	90	(13)	—	—	—	—	—	—	—	—	90
Total	700	(100)	637	(100)	166	(100)	340	(100)	253	(100)	2,096

Note: Based on data obtained from *Chemical Facts*, various years.

chemical subsidiaries had substantial capacities in both LDPE and PS.

Following the oil price rises, the structure of the industry started to change, when subsidiaries of oil companies were increasing their share of the market at the expense of traditional producers, particularly in PVC, PE and PP (see 1980 capacities), while American chemical subsidiaries held their strong position in PS.

The companies based in the United Kingdom had improved their position to what it had been in the mid-1960s when in 1980 they owned over 80 per cent of the thermoplastic production capacity. The European chemical producers' participation in the British production capacity was only minimal relative to the role played by the American companies.

The trend during this period was for fewer companies to concentrate their production capacities on certain product areas, while smaller producers were being forced to drop out of the market.[28] As larger and larger capacity plants were coming on stream, costs for plants and machinery were escalating, so that, new plants required increasing capital commitments.

3.7. Structural changes in the British petrochemical industry

The additional capacities that were coming on stream, following the growth recovery after 1975, coincided with sluggish markets in 1980, when production was declining and prices stagnating as competition increased, with the overall effect of squeezing profit margins.

Prices of 'maturing' products such as PVC and LDPE have a tendency to decline as the number of producers and demand increase. Production costs were reduced as larger plants achieved economies of scale. Producing firms also benefitted from the 'learning curve' process when improved production and marketing techniques were developed. Another factor affecting prices is that of *overcapacity*; firms running older plants accept lower profits by selling at low prices after having achieved an acceptable return on investment for their existing plants.

The British petrochemical industry witnessed the closure of at least twelve petrochemical plants[29] over the period 1981–2, when the industry in its usual manner attributed this to:

1. Overcapacity, where plants were running at 60 to 70 per cent of full capacity;
2. A decline in profits, resulting from lower margins, since product prices were under pressure while feedstock prices were still increasing.
3. Demand for petrochemical products, particularly plastics, declined after 1979 as a result of the slump in the production of major purchasing industries, particularly in construction and motor vehicles.

While it is clear that the demand for plastics materials in the United Kingdom has been depressed throughout this period, there is no evidence that prices were falling in real terms. The wholesale price indices of most plastics materials in 1978 had shown a considerable increase over the 1976 prices, which on average was about 41.4 per cent for synthetic resins (see Table C2).

On the other hand, the problem of overcapacity is not something new to the industry; it had arisen even during the fifties and sixties. As early as 1931, the president of the Society of the Chemical Industry identified the problems facing the industry at that time as the following (Sharp and West, 1982): falling in prices; overcapacity; increasing competition from Japan and the developing countries; economic nationalism and unfair trading practices.

Overcapacity is characteristic of the petrochemical industry, where production capacity increases in large steps, especially with modern large-scale plants coming on stream, whereas demand increases at a modest step-wise pace.

Despite complaints about overcapacity, low prices and low profit margins, capacities were still being added in 1980 and more plants were being planned or were under construction. Shell added a further 70 thousand tonnes/year of LDPE to its existing capacity at Carring. British Industrial Plastics extended its PVC capacity by 50 thousand tonnes/year at Aycliffe. ICI added further extensions to its polyester resin and ethylene glycol units at Wilton. BP added 54 thousand tonnes/year of HDPE and 93 thousand tonnes/year of PVC.

More plants were still under construction or under consideration in 1982, some of which were postponed from 1980. These include PP, PVC and PS units of 50, 150 and 20 thousand tonnes/year respectively for ICI.

The British petrochemical producers closed a number of plants permanently and several plants changed hands. These moves can at best be viewed as serving the following purposes:

1. Reshape the industry by closing down old plants,[30] while new investments were planned for plants, employing new technologies using North Sea gas-based feedstocks, which are more economic than naphtha-based plants. Also in anticipation that more of the same products would be available at competitive prices from Canada, Mexico and the Middle East by the end of the decade.
2. The British deals meant that BP would pull out of the PVC market by closing down six of its old plants (Quinland, 1982) and transferring the remaining one to ICI. In exchange ICI pulled out of the LDPE market and increased its share in the joint ICI-BP ethylene cracker, thus leaving each company with the product in which it had a comparative production advantage (BP with its access to ethane and ICI with its firm hold on chlorine) and a large market share. Excess capacity was curtailed and better control achieved of prices, which are normally set by the producer with the largest market share, forcing other producers to follow suit.[31]
3. Oil companies were moving into the oil-processing business, while major companies were taking appropriate measures that would allow them to phase out their less profitable operations and move into more profitable markets with better growth prospects such as speciality products, for example, special grade polymers, surfactants, pesticides, pharmaceuticals, and so on.
4. The major petrochemical producers in the United Kingdom have international investments in various regions of the world, the United States, Western Europe and the Far East. The rationalisation of their British capacity would be part of their plans to move into markets where higher growth rates are expected or where production costs are more favourable than export supplies from their home-based production units. Many companies have reduced their operations in Western Europe and moved into the American market.

Figure 3.10
British share of Western European polyethylene production

Figure 3.11
British share of Western European PVC production

Figure 3.12
British share of Western European polystyrene production

Figure 3.13
British share of Western European amino plastics production

Figure 3.10

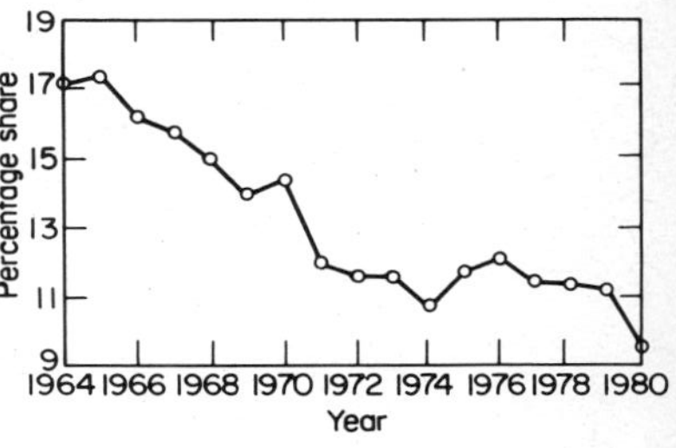
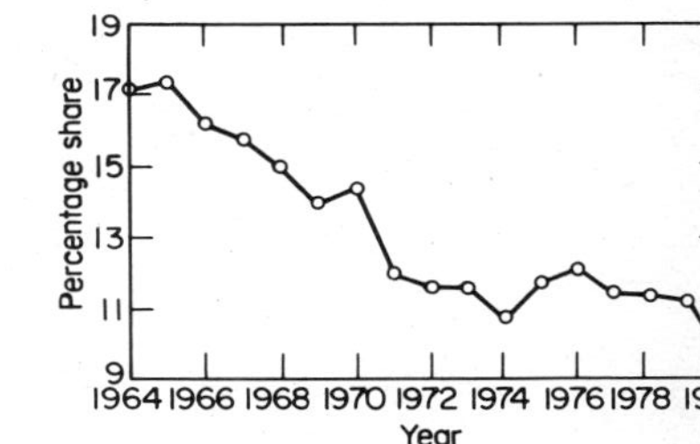

Figure 3.11

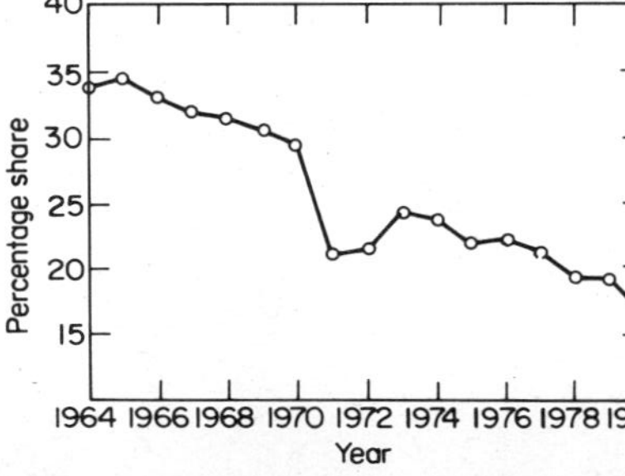
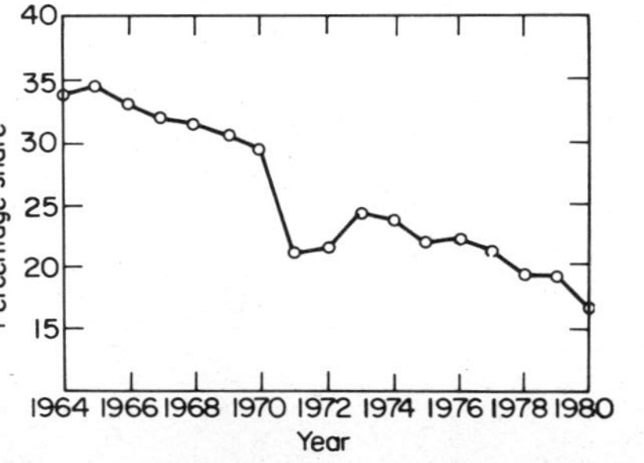

Figure 3.12

Figure 3.13

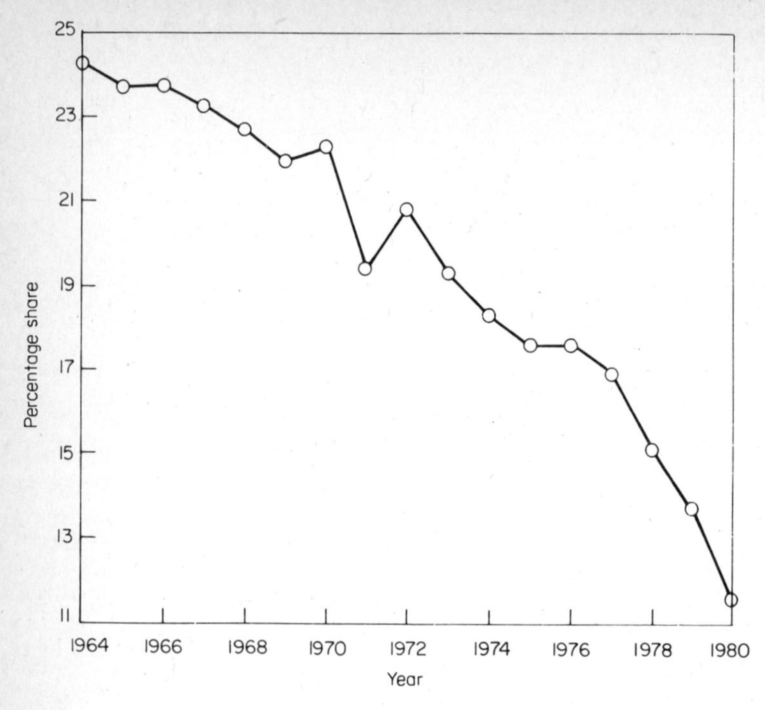

Figure 3.14 British share of Western European synthetic rubber production

3.8. The trade in petrochemicals and plastic materials: the British position

The British positive trade balance in plastics materials declined after 1968, becoming negative in 1978. With the EEC, it has been negative since 1973 and continued to deteriorate in their favour into 1980 (see Figures 3.10–3.14). However, the British overall trade in petrochemicals, including bulk organic chemicals, with the rest of the world was still positive, although the imports into the British market had continued to increase for both organic chemicals and plastics materials (NEDO, 1981), rising from 33 to 34 per cent for organic chemicals and plastics materials respectively in 1976 to about 45 to 41 per cent in 1980.

This situation has been largely a result of the combination of the following factors:

1. Shortages in supplies and price escalation of natural gas as a source of energy and feedstock;
2. Inconsistencies in government policy on the issue of pricing energy sources, which has resulted in
3. A climate of uncertainty for future investments in addition to the record of slow commissioning of large plants in the United Kingdom.

Together these factors have an adverse effect on the petrochemical industry of this country.

3.9. Energy, pricing and supplies to the chemical industry in the United Kingdom

The British petrochemical industry pays higher prices than its European competitors for its feedstocks and energy which, furthermore, are less readily available in the United Kingdom.

Table C3 shows that the British prices for natural gas are higher than those for the rest of the Community, with the exception of West Germany. The average unit price increases over the level of prices prevailing in 1973 which were also highest for the United Kingdom and Italy, where these were, respectively, 390 and 460 per cent of the 1973 prices.

From 1977 onwards, heavy fuel oil was also priced higher in the United Kingdom and West Germany than in the rest of the EEC (see Table C5). The average price increases over the level of 1973 were again significantly higher for the United Kingdom and Italy; even in 1982, when the United Kingdom was one of the major oil-producing countries, fuel oil prices for the process industries were about 10 per cent higher than in the rest of the EEC (Caudle, 1983).

However, until 1982 the largest disparity in energy prices was in industrial electricity. The British chemical industry is one of the largest electricity consumers, accounting for about 20 per cent of the total electricity used by manufacturing industries. Yet, the Electricity Council prices (Caudle, 1983) for high-load factor (10 megawatt) industrial consumers were about 25 per cent higher in the United Kingdom than in West Germany, 40 per cent higher than in France and 50 per cent higher than in Italy; they were even greater for larger users (40 megawatt).

For most commodity products (plastics, particularly LDPE and PVC), the electricity content, in the United Kingdom, represents 15–30 per cent of the product prices. Such disparities in energy prices are detrimental for an energy-intensive industry competing with neighbouring producers that, although they have no advantage in natural energy resources, enjoy more favourable taxing schemes, and are supported by government efforts to supply energy to their major industries at low costs.

The case of the British chemical industry has been brought into discussion with the Government recently at National Economy Development Council (NEDC) meetings (NEDC, 1980), however, the outcome of these discussions and the constructive measures that are needed to be taken have yet to be seen!

3.10. The construction and the car industries in the United Kingdom

The demand for, and consequently the production of, plastics materials in the United Kingdom was badly hit in 1979–82 period by the recession in general and in particular as a result of the decline in the output and investment in two important industries accounting for a major part of the final demand for plastics materials, namely the construction and the motor vehicles industries.

The gross domestic fixed capital formation in new buildings and works was declining from 1975 onwards, until in 1980 the total investment for all sectors was only 85 per cent of the 1975 level, which was a result of reduced government spending on this sector, as can be seen from Table C5. Particularly the housing sector (Table C6) suffered a serious decline in total investment, which by 1980 had declined by 25 per cent from its 1975 level, and this was again largely a result of declining investment in housing by the public sector.

The prices of finished plastic building materials, however, continued to rise in the United Kingdom despite the depressed markets and the downward pressure on prices for commodity petrochemical products during the 1978–81 period. The percentage price increases of plastic building materials relative to their 1975 prices were higher than those of metal building materials products (see Table C7).

Table C8 gives the values of imports and exports for building materials over the period 1975–80. The data show that in 1980, while exports of plastics building materials and electric wires (an indirect outlet for plastics materials) still exceeded imports. There was, since 1975, a declining trend in the ratio of exports to imports as a result of increased import penetration in these two types of products, where the import/export ratio increased from 0.14 in 1975 to 0.40 in 1980 for electrical wires and from 0.51 to 0.68 for plastics building materials over the same period, indicating an average annual increase in the level of import penetration of 37 per cent for electrical wires and 6.7 per cent for plastics building materials.

The other major user of plastics and synthetic materials is the car industry, whose output is largely influenced by personal consumption and is an important final demand market for plastics materials. The production of passenger cars in the United Kingdom has been declining continuously since 1972, when production was 1.92 million units which had fallen by 1 million units by 1980, when total production was back at the level of 1957 of 0.92 million units. The fate of the European car producers over the same period was much better with production increasing continuously in both France and West Germany, which stood at a production peak, in 1979, of 3.72 million and 3.94 million cars respectively.[32]

The consumption of plastics materials per unit car was on the increase during this period, when heavy metal parts were being substituted wherever possible to reduce the average car weight and increase its mileage/gallon. Again, the case was that United Kingdom car manufacturers were slower in introducing these changes than their European counterparts.

3.11. Concluding remarks

Using an input–output approach to estimate the structure of demand for plastics materials, it may be concluded that industrial output is a far better indicator of the level of demands for petrochemical products than GDP, since most of the output of plastics materials and petrochemical products is consumed by a small number of other industries in the form of intermediate inputs to these industries.

It has also been established that input–output models offer an accounting technique of the whole economy, which covers the inter-industry transactions and the effects of final demands, where the matrix of technical coefficients displays the structure of the economy and allows the investigator to analyse the growth of particular industrial sectors and to evaluate the impact of these changes, together with those of final demand components on the economy as a whole.

While certain industrial sectors may show rapid structural, technological or growth rate changes over short or medium periods, the economy as a whole changes only gradually in a well-defined manner, making it possible for input–output models to forecast, with a good deal of reliability, the growth of industrial outputs.

Input–output models are, therefore, useful as a planning tool or as a means of forecasting demands for petrochemical products and projecting their outputs. The use of input–output models in this country has so far been very limited. It is expected, however, to increase, as the Central Statistical Office improves and expands its data collection and as the usefulness of input–output models is drawn to the attention of management and planning departments of major industrial firms.

On the other hand, the review of the supply side of petrochemical products in the United Kingdom reveals that the production of the bulk of these products is controlled by a small number of companies (as is the case in other major producing countries), with ICI and BP being involved in the production of a wide range of products and having the largest production capacities. Together with Shell, these three companies control between 60 per cent and over 90 per cent of the production of most products.

The expectations of high growth rates, of the same order as those of the 1960s, for petrochemical products led many companies to invest heavily or expand their production capacities after the first oil price rise and the recovery of demand in 1976. However, this surge in demand was short-lived, as European countries and the rest of the world were hit by a deep recession. In the face of excess capacity and declining demand, the competition in the European market increased, especially in the United Kingdom, as companies tried to increase their sales volumes which put prices under pressure.

Recent developments show that larger companies were tightening their grip over the market, with many oil companies penetrating the petrochemical market with substantial capacities of production.

The stagnation in the economic activity and the decline of industrial output in the United Kingdom were the general factors depressing the demand for plastics materials, but in particular this can be attributed to the industries which are the major users of plastics materials whose outputs were declining rapidly after 1979, such as the motor vehicles and the building and construction industries.

Exports of plastics and synthetic resins have also been affected badly by the following factors which occurred at different times:

- the high rate of exchange of the sterling pound during 1979–81;
- the higher costs of energy and utilities for the United Kingdom industry relative to other European producers;
- increased competition from overseas producers, especially those with cheaper feedstocks and energy supplies, such as the United States, Italy and Canada;
- many of the importing countries had started producing petrochemical products in their home markets.

4 THE PETROCHEMICAL INDUSTRY IN DEVELOPING COUNTRIES

4.1. Introduction

In Chapter 2, the changing pattern of the production cost structure of petrochemical products and the changing roles of the major producers, resulting from the rising oil prices and the increasing maturity of the industry, were discussed. In this chapter, the impact of these developments on the emerging petrochemical sector in the developing countries is dealt with. Particular attention is given to the development plans of the major oil producers of the Middle East, as these involve large-scale export-oriented petrochemical plants throughout the region.

The production capacities and the consumption patterns of various products in the developing countries are studied. The production costs are analysed and compared with those of the major producers of Western Europe, Japan and the United States to help in assessing the possibility of viable exports into these latter regions. The difficulties facing the new producers and the benefits that will accrue to them from promoting these projects in a substantial way are also discussed.

4.2. The advent of petrochemicals in the developing countries

Until the early 1970s, production facilities for downstream petrochemicals in almost all developing countries were virtually non-existent. Their relatively small demands for synthetic materials, namely plastics and fibres, were met by imports from the major petrochemical producers of Western Europe, the United States and Japan.

In 1975, demand for the major plastics materials (LDPE, PVC, HDPE, PP and PS) in developing countries was 3.3 million tonnes (Table 4.8 Chapter 4, Section 4.5.2), with most of this demand being concentrated in the Latin American region (1.1 million tonnes) and Asia (1.2 million tonnes), whereas the Middle East and North Africa together accounted for about half a million tonnes.

Nevertheless, the 1970s was a period of active development in the field of petrochemicals in the developing countries, where the technology for the production of basic petrochemicals and the major plastics materials was becoming mature and widely available from engineering firms and oil companies (imitators). Also, the oil price rises had altered the production cost structure of these products in favour of feedstock and raw materials, which ultimately accounted for about 80 per cent of the total production cost (excluding ROI; see Chapter 2).

The process of internationalising the petrochemical industry had already been under way by the late 1970s, with the developing countries increasing their share though still very small, of the world petrochemical output. Countries with adequate refining capacity and expanding markets found an opportunity to produce their petrochemical needs locally, expand their industrial output and replace imports. However, the roles of the different developing countries vary in importance according to their industrial activity and feedstock potential. Two groups of operators, which are very active among the developing countries in the field of petrochemicals, can be identified: the oil-producing countries, namely the OPEC members, and the newly industrialised countries (NICs).

The roles of these two groups will be discussed in the following section, which looks closely at their potential as petrochemical producers and the future of their ambitious plans.

4.2.1. The oil-producing developing countries (OPDC)

In the 1960s and early 1970s, hardly any member of the OPDC was involved in the production of petrochemicals with the exception of ammonia and urea, despite the abundance of cheap raw materials and potential markets. The oil price rise of 1973 changed this and also gave the major oil-exporting countries an immense financial capability which they sought to utilise in improving their poor infrastructure and building an industrial base around the oil sector.

The Middle East region is endowed with about 60 per cent of the known world oil reserves, which the regional planners saw as an important asset for entering the petrochemical industry even before the oil price rise. As early as 1970, Baghir Mostafi, the director of National Iranian Oil Company declared: 'The era in which the petrochemical resources of the region were permitted to remain dormant is drawing to a close' (Anderson, 1970). His views were, no doubt, shared by those in similar positions in the oil-producing countries of the region. Many similar statements have appeared since then. However, the plans for petrochemical ventures were taken more seriously by the second half of the 1970s. Most of the petrochemical plants in the region use associated gas as a feedstock which was flared at an estimated rate of 100 billion cubic metres per year (equivalent to about 2 million barrels of oil per day in the oil-producing countries of the Middle East (UNIDO, 1979)).

The oil-producing countries of the Middle East were the most forthcoming among the OPEC members in announcing their intentions to become major producers of petrochemicals in the early 1970s. They had ambitious plans for more than seventy major refineries and petrochemical plants. However, many of these were either abandoned or shelved subsequently owing to the depressed situation in the petrochemical industry. The overcapacity in most basic petrochemical products, especially in Western Europe, and also the weak demand for petrochemicals since the late 1970s forced these countries to postpone their plans.

Until 1975, only Algeria, Indonesia, Iran and Venezuela of the OPEC members had operational petrochemical capacities and these were only in two basic products, ethylene and ammonia (see Table D1), whereas the remaining member countries with much larger feedstock reserves were still at the planning stage of their petrochemical ventures. By 1980, many of the planned capacities of world-size plants were already in operation in most member countries. Only Algeria and Libya had operating methanol plants, but none were involved in BTX production. More of the planned capacities are expected to come on stream by 1985–6, notably the large export-oriented plants that are being built in Saudi Arabia, Indonesia, Iran, Kuwait and Libya.

Saudi Arabia and Iran are involved in plans for the production of a full range of basic petrochemical products, including ethylene and its derivatives, styrene, ammonia and urea. Iraq, Qatar and Algeria are already involved in the production of some of these products,

Table 4.1 Major petrochemical projects in the Middle East

Project/country	*Company*	*Capacity (tonnes/year)*
Ethylene crackers		
Iran	Iran/Japan Petrochem.	300,000
Iraq	Ministry of Industry and Mining	150,000
Kuwait	Petrochem Ind.	300,000
Qatar	Qatar Petrochem./CdF. Chimie.	250,000
Saudi Arabia	Sabic/Shell	650,000
Saudi Arabia	Sabic/Mobil	450,000
Saudi Arabia	Sabic/Dow*	650,000
HDPE		
Iran	Iran/Japan Petrochem.	60,000
Iraq	Ministry of Industry and Mining	30,000
Qatar	Qatar Petrochem./cdF. Chimie	90,000
Saudi Arabia	Sabic/Mobil	70,000
Saudi Arabia	Sabic/Dow*	80,000
LDPE		
Iran	Iran/Japan Petrochem.	100,000
Iraq	Ministry of Industry and Mining	60,000
Kuwait	Petrochem. Ind.	130,000
Qatar	Qatar Petrochem./CdF. Chimie	140,000
Saudi Arabia	Sabic/Exxon	260,000
Saudi Arabia	Sabic/Dow*	70,000
Saudi Arabia	Sabic/Mitsubishi	130,000
Saudi Arabia	Sabic/Mobil	200,000
EDC		
Iran	Iran/Japan Petrochem.	300,000
Saudi Arabia	Sabic/Shell	450,000
PVC		
Iraq	Ministry of Industry and Mining	60,000
Methanol		
Bahrain	Gulf Petrochemicals	330,000
Saudi Arabia	Sabic/Japan Petrochem.	660,000
Saudi Arabia	Petromin/Borcdcrn/Houston NG	500,000
Saudi Arabia	Sabic/Celanese/Texas Eastern	650,000
Styrene		
Kuwait	Petrochem. Ind.	334,000
Saudi Arabia	Sabic/Mobil	320,000
Saudi Arabia	Sabic/Shell	295,000

* Dow withdrew from the partnership with SABIC in late 1982. SABIC, however, decided to go ahead with the project as planned.

Source: Based on data in *Hydrocarbon Processing,* February and June issues, 1981–4.

while Kuwaiti units for ethylene, LDPE, styrene and aromatics are still at the planning stage.

The largest of these plants are those in Saudi Arabia, where three world-scale ethylene crackers with a total capacity of 1.6 million tonnes are expected to come on stream in 1985/86 (two actually started in 1985) with substantial amounts of this output going to the production of downstream products such as LDPE, HDPE and VCM. Urea, ammonia and methanol would also be produced in very large quantities, destined for export markets (see Table 4.1).

Iran, on the other hand, has had under construction large-scale petrochemical complexes which were due to have come on stream by the early 1980s – Iran's petrochemical capacities, like those of Algeria and Iraq, were planned to meet their domestic demand for these products which would supply their newly formed industries and the expanding home market, with the surpluses finding their way to international markets. However, the constructions here have all been disrupted by the Iran–Iraq war and the acute cash shortage. Some of the large-scale plants under construction in Iran were planned on a joint-venture basis with Japanese participation.

The less populated countries of Kuwait, Qatar and Libya are in a similar position to Saudi Arabia, with large planned capacities aimed at export markets since the local demand for petrochemicals in these countries is very small.

All Middle East oil-producing countries had large production capacities for ammonia and urea, which were among the first products to be produced in the area, and have already established their markets in developed and developing countries. In 1980, these countries had a combined output of 5 million tonnes of ammonia, which accounted for about 6.5 per cent of the total world ammonia output. Algeria, Iraq, Iran, Kuwait and Qatar were the largest producers in the region with additional capacities expected to come on stream in the second half of the 1980s in Saudi Arabia and other producing countries, whose outputs would go for exports as well as to meet the needs of the large agricultural projects being developed in the area.

In aromatics, only Iran, Kuwait and Saudi Arabia have plans for the production of large quantities of BTX and styrene for export markets. Plants, each of half a million tonnes capacity for BTX, are expected to come on stream in Iran and Kuwait in the second half of the 1980s, while Saudi Arabia and Kuwait each will have a capacity of 300,000 tonnes of styrene by that time.

Overall, the Middle East capacities of petrochemicals would still constitute only a small percentage of the total world capacity by 1990, ranging between 5 per cent for ethylene to about 11 per cent for each of ammonia and methanol. But these countries' involvement would constitute a large share of the new world capacities under construction, which are expected to come on stream during the 1980s (OPEC, 1981).

Other oil-producing developing countries were also involved in planning and producing petrochemicals. Indonesia and Venezuela are among the leading oil producers which already have substantial producing capacities and are planning to expand further these capacities. While Nigeria, Ecuador and Gabon are expected to become petrochemical producers in the late 1980s, when their first petrochemical capacities come on stream. A new development in the petrochemical industry at this stage was the arrangement for OPDC-producing countries to supply significant amounts of their petrochemical output to member countries of their regional markets. Ecuador and Venezuela would supply between 30–50 per cent of their ethylene, urea, HDPE, PP and PS outputs to the Andean Pact market, while Indonesia has a joint venture for the production of urea and ammonia with its Asian members. Co-operation among the Middle East producers, however, has been very low key, but efforts are being made to improve the situation through the formation of various regional development and economic agencies.

4.2.2. The newly industrialised countries (NICs)

This group of countries includes the developing countries which have a substantial and rapidly expanding industrial base, such as India, South Korea, Brazil, Mexico and Argentina. Only Mexico, among the NICs, has a large reserve of oil and gas and is a major exporter of oil.

The Latin American members of this group were involved in petrochemical production, on a small scale, since the early 1960s. Today these countries have production capacities for a wide range of petrochemicals, including basic products, plastics, synthetic fibres and rubbers and detergents.

As they do for other industries, these countries gave protection to their petrochemical industry in its early stages of development by erecting high tariff barriers against imported products and by encouraging joint ventures with foreign petrochemical companies in the downstream products. However, the control of the basic producing units, such as ethylene steam crackers, is totally in the hands of the state-owned companies. The sufficiently large home markets, the expanding industrial output and the well-established infrastructure of these countries contributed to the rapid expansion of their petrochemical output. By the mid-1970s, these countries accounted for most of the production of petrochemicals in developing countries. Brazil and South Korea have no oil or gas resources of their own, yet they had substantial petrochemical production capacities and were the largest petrochemical producers among the developing countries. They also accounted for a large amount of exports to neighbouring countries, while the major oil-producing countries were still in the early stages of developing their petrochemical-producing capacities.

The developing countries, particularly the oil producers, were hoping that their petrochemical investments would help in diversifying their income sources and generate more value added from their oil resources, increase their exports and save hard currency by substituting petrochemical imports. The development of a petrochemical industry was also seen as a driving force for expanding the home consumer market as well as supplying the intermediate products to other industries such as textiles, plastics processing, fertilisers and building materials. For the more industrially advanced of these countries (Brazil, Mexico, South Korea, India), the petrochemical industry would help them develop an engineering fabrication sector, increase know-how and develop an indigenous scientific base in a national process industry.

4.2.3. The roles of the chemical and oil majors in joint ventures in OPDC

The recent developments in the petrochemical industry in developing countries, in particular the export-oriented plants in the oil-producing countries, are of major concern to producers in industrialised countries, mainly in Western Europe, who see this as a challenge to their supremacy in the export field, which also threatens their national industries and markets if the oil-producing countries were to go ahead with their plans in a big way.

The major joint ventures were arranged in the Middle East, following the 1973 oil price rise, between the largest oil and chemical producers and the leading oil-producing countries. The most significant of these ventures were planned for Saudi Arabia and Iran (see Table 4.1) for the production of basic petrochemicals such as ethylene, methanol and ammonia as well as plastics polymers (LDPE, HDPE, etc.).

It must be noted, however, that the involvement of the oil companies was wider than that of the chemical majors, since the former found an opportunity to expand their petrochemical base and obtain oil entitlements offered by the oil producers in return for their participation in these joint ventures. Also, the oil companies were able to develop their own process technology for such products and were more suited to deal with the oil producers than the chemical majors, drawing from their past experience and historical presence in the area.

Shell, Exxon and Mobil (the smallest member of the Aramco consortium) were among the leading oil companies involved in the joint ventures in Saudi Arabia; while the chemical majors included Dow, Celanese and Mitsubishi. Mitsui and Mitsubishi were also involved in joint ventures with Iran, which were under consideration since the early 1970s. The French firm CdF-Chimie had a minority share in Qatar's ethylene and polyethylene (LDPE; and HDPE in the future) plant, which came on stream in 1980.

On the other hand, the major chemical companies were less forthcoming in their involvement in these projects since by now these companies were moving away from bulk petrochemicals into more profitable speciality products. While those involved found an opportunity to obtain cheap energy resources for their ventures and to secure their feedstocks in general. It is significant that the Japanese companies, which have no hydrocarbon resources at home, were very active in these developments, receiving considerable backing from their government as their involvement gave them direct access to the world's most resourceful region in oil and gas. The Japanese government also saw their companies' intake of bulk petrochemical products as an indirect and safe way of importing energy. The Japanese petrochemical companies were also involved in joint ventures in South Korea (Mitsui petrochemical) and Singapore (Sumito chemical) for the production of ethylene, LDPE, HDPE, PP and ethylene glycol.[33] The French state-owned CdF-Chimie was in a similar position, as a company short of access to oil.

In their turn, the Gulf oil producers, by forming these joint ventures in which the foreign partner was involved with equity shares and substantial financial investment (50 per cent in the case of Saudi Arabia), were trying to overcome their lack of experience and capability to operate and manage these large-scale plants as well as ensuring an outlet for their outputs through the marketing networks of the joint-venture partner.

The chemical engineering firms were very active during this period in spreading petrochemical capacities throughout the world, building chemical plants for which they had developed their own process technology or obtained under licence from innovating chemical companies. In particular, these firms were capable of developing their own processes for the production of basic petrochemicals. The following engineering companies, the majority of whom are American-based, Fluor, Bechtel, Foster Wheeler, Lummus and Kellogg (US), Chiyoda (Japan), Snamprogetti and ENI (Italy), Technip (France) and Uhde (Germany), accounted for the construction of most of the ethylene, methanol and urea plants that came on stream around the world since the mid-1970s, using their own technology. Their involvement in petrochemicals, independently of the major chemical and oil companies, was mainly in supplying turn-key projects to the NICs and East European countries. In the Middle East they were very active in Algeria, Libya and Iraq. Their future role will continue to increase in importance in this area of petrochemicals, as can be judged by their involvement in the plants that are still at the planning stage.

This very brief assessment of the roles of main petrochemical operators (chemical majors, oil majors and chemical engineering construction firms) agrees completely with the findings of Stobaugh (1968) in a study on petrochemicals, where he finds that for maturing products the imitator producers would increase their production involvement at the expense of the original innovators in a pattern as shown in Figure 4.1, and where production under licence rises dramatically at this stage of the product life cycle, thus contributing to rapid increase in world production of these mature products.

4.3 The changing structure of petrochemical production costs: the implications for OPDC

In Chapter 2, the changing structure of production costs of

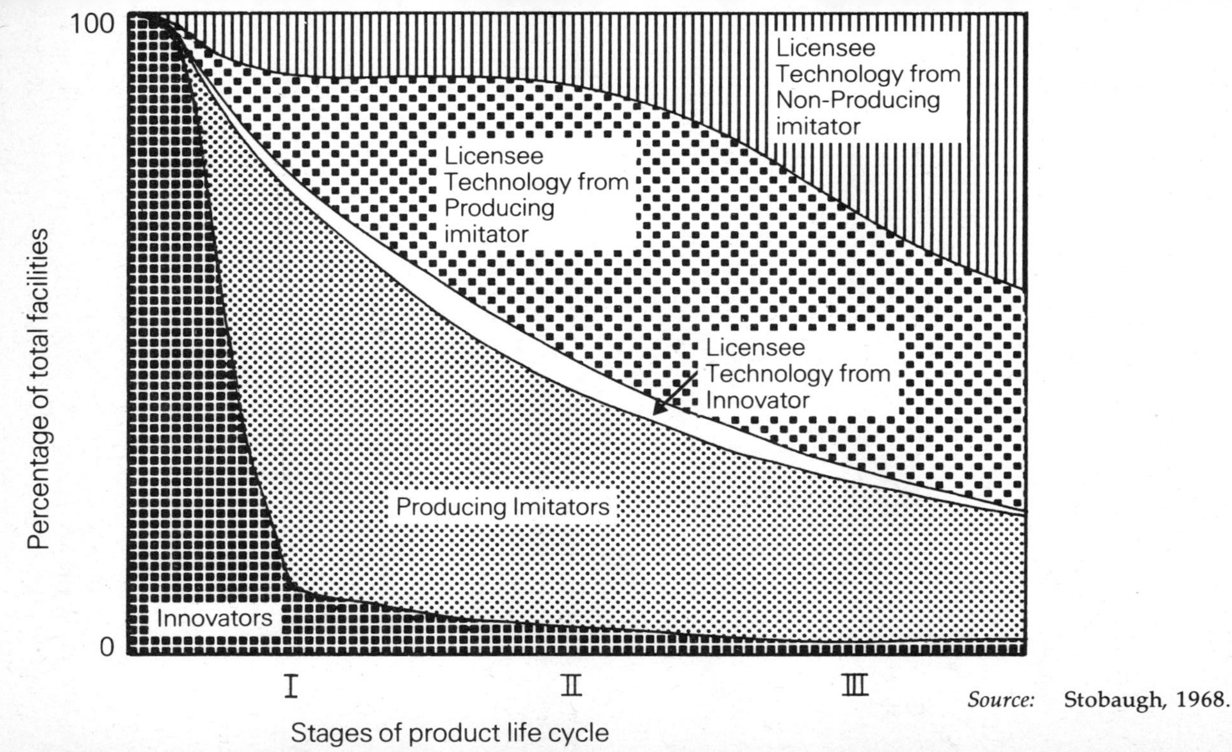

Source: Stobaugh, 1968.

Figure 4.1 Share of world production facilities built during each stage of product life cycle owned by categories of firms

petrochemical products was assessed for the conditions prevailing in Western Europe and the United States, taking into consideration the effects of various plant sizes and the corresponding fixed costs of these plants on the production costs over the life span of plants of various ages, i.e. built at different periods.

In this section, however, the production costs for a number of petrochemical products will be looked at in the context of conditions prevailing in industrialised major petrochemical producing countries and compared to those of developing oil-producing countries (OPDC) with cheap energy sources. The implications of these comparisons will be assessed in the light of future developments of increasing world demand for petrochemicals and rising costs of energy.

The oil price rises of the 1970s have effectively changed the production cost structure of many petrochemical products in favour of the feedstock costs, which eventually accounted for between 50 per cent and more than 75 per cent of the total production costs (including a 25 per cent ROI and 10 per cent depreciation charges) for basic petrochemicals, and were between 35 per cent to 55 per cent for downstream products (see Chapter 2).

The implication of the new situation, where the production cost structure of most petrochemical products has become dominated by the feedstock costs, is that OPDC found an opportunity to produce competitively a wide range of petrochemical products by utilising some of the vast quantities of associated gas (the gas produced in association with oil production), of which they have been flaring between 40–60 per cent (OPEC, 1981).

The OPDC, especially the major oil producers of the Middle East, have access to vast quantities of oil and associated gas at very low cost. The ethane feedstock, around which most of the export-orientated petrochemical plants are being built in these countries (Table 4.2), would be supplied to these plants at $25 per tonne, which is the total cost of separating it from the gathered associated gas, since the vast quantities of this gas have no other use in these countries except perhaps for export as LPG. In Japan and Western Europe, naphtha is the main feedstock for the industry. It is, therefore, clear that the OPDC have a considerable feedstock advantage over the major petrochemical producers of the world.

Furthermore, the associated gas of the Middle East is very rich in ethane and propane (see Table D2) making it an ideal feedstock for

Table 4.2 Gas-based production of petrochemicals in some developing countries (existing plants plus planned new capacities) tonnes/year

	Ethylene	*LDPE*	*HDPE*	*EDC/VCM or PVC*	*Styrene*	*Ammonia*	*Urea*	*Methanol*
Algeria	140,000	48,000	—	35,000	—	991,000	278,000	100,000
Bahrain	—	—	—	—	—	660,000	—	350,000
Indonesia	300,000	185,000	60,000	110,000	—	2,173,000	3,954,000	400,000
Iran	325,000	100,000	60,000	40,000	30,000	1,106,000	1,180,000	—
Iraq	130,000	60,000	30,000	60,000	—	994,000	1,535,000	135,000
Kuwait	350,000	130,000	—	—	320,000	994,000	1,360,000	—
Libya	—	—	—	—	—	663,000	900,000	330,000
Malaysia	1,945,000	—	—	—	—	380,000	328,000	550,000
Mexico	300,000	499,000	200,000	570,000	333,000	4,796,000	1,691,000	1,007,000
Nigeria	—	120,000	60,000	120,000	—	330,000	496,000	—
Qatar	280,000	140,000	70,000	—	—	595,000	991,000	—
Saudi Arabia	1,606,000	640,000	171,000	454,000	295,000	528,000	500,000	1,320,000
Syria	—	—	—	—	—	300,000	346,000	—
Trinidad and Tobago	—	—	—	—	—	1,280,000	70,000	435,000
UAE	—	—	—	—	—	330,000	—	—
Venezuela	150,000	50,000	60,000	54,000	—	792,000	661,000	—
Total	5,914,000	1,705,000	642,000	1,443,000	978,000	16,992,000	14,790,000	4,627,000

Source: UNIDO and GOIC 1981.

petrochemical production. Gas-based ethylene plants also have the advantage of not producing the different by-products associated with the use of naphtha feedstocks, thus saving the OPDC the problem of marketing these by-products that may not find a direct industrial use in their markets.

However, the other determinant of production costs for petrochemicals is that of capital investment costs, which continued to rise well over the rate of inflation in the 1970s and are expected to continue rising at 2 per cent in real terms in the 1980s.[34]

In general, capital investment costs are much higher in developing countries than in the United States and Japan, where they are lowest. In particular, the smaller oil-producing countries of the Middle East and Indonesia are at a disadvantage in this respect because investment costs there are between 50–110 per cent higher than on the American Gulf Coast (taken as a basis for international comparisons in the oil-processing industry) for plants of the same size.

These higher costs result from a combination of the following factors: the delays in executing projects, insufficient supply of indigenous skilled manpower, almost complete lack of engineering services, all manufactured materials and equipment having to be imported from distant places and assembled under unfavourable weather conditions and, finally, an inefficient infrastructure in general. Usually engineering construction firms assess the combination of these factors as a single 'location factor', which they use in their calculations for investment costs in various locations around the world. Table D3 shows the disadvantaged position of the OPDC, where investment costs per ton of product for various petrochemicals are higher than those of the United States, Japan and West Germany in accordance with their location factors, which range from 1.25 for Mexico to 2.1 for Indonesia. The differentials in these unit investment costs were highest for downstream products such as ethylene glycol, ethylene oxide, LDPE, PP and PET.

The higher investment costs incurred by the OPDC offset the feedstock advantage enjoyed by these countries and led to a situation where their competitiveness would be limited to basic products such as ethylene, ammonia, urea, methanol, HDPE, styrene and VCM, whose production cost is heavily dominated by the feedstock component.

However, it can be argued that the OPDC would be more

Table 4.3 Calculated production cost for selected petrochemicals ($/tonne – 1980)*

	USA Gulf Coast	*FRG*	*Japan*	*Indonesia*	*Mexico*	*Qatar*
Feedstock price calculated to include:	*25% ROI*	*25% ROI*	*25% ROI*	*5% ROI*	*5% ROI*	*5% ROI*
Ammonia[†]	317	345	375	195	126	151
DMT	1,265	1,417	1,574	1,178	842	928
Ethyl benzene	782	978	1,168	680	556	618
Ethylene[‡]	630	—	—	375	282	290
Ethylene-propylene[§]	613	—	—	437	315	360
Ethylene-propylene-butadiene/benzine[‖]	733	948	746	—		
Ethylene glycol	739	919	1,053	1,107	708	846
Ethylene oxide	965	1,282	1,287	905	581	695
HDPE	1,061	1,380	1,479	886	625	737
LDPE	979	1,295	1,367	849	560	658
LLDPE	951	1,243	1,311	848	606	710
Methanol	288	313	352	136	93	111
PET[¶]	1,773	1,808	2,157	2,592	1,759	2,235
PP	986	1,129	1,283	1,112	727	944
PS	1,068	1,262	1,474	1,185	774	1,051
PVC	1,090	1,311	1,473	1,699	796	1,343
SBR	2,079	2,286	2,335	1,856	1,255	1,671
Styrene	893	1,069	1,231	938	604	831
TPA	1,207	1,381	1,389	1,201	876	972
Urea	169	197	349	168	109	134
VCM	798	996	1,048	902	647	639

* At 85 per cent load feedstock.
† Methane feedstock at current market price.
‡ Ethylene production cost with ethane feedstock at current market price.
§ Ethylene production cost with ethane-propane feedstock at current market price.
‖ Ethylene production cost with naphtha feedstock at current market price.
¶ DMT feedstock.

Source: UNIDO 1981.

competitive than the traditional industrially-developed producers of petrochemicals in the production of a wide range of products if some of the stringent assumptions, that are usually used in calculating the production costs (such as ROI at 25 per cent, depreciation at 10 per cent, working capital at 20 per cent of fixed capital costs, relatively high interest rates on investment) of petrochemicals, were relaxed. For instance, if Mexico, Qatar and Indonesia were to accept a 5 per cent ROI, then they would be able to produce more competitively than their neighbouring developed countries a wide range of downstream products such as ethylene glycol, LDPE, PVC, PS, PP, SBR etc., despite their higher investment costs as indicated in Table 4.3.

While it is widely accepted that the investment costs in OPDC are at present significantly higher than those of major petrochemical producers, one would question the accuracy with which the values of the 'location factors' are arrived at. It is also questionable if the assumptions covering the estimation of ROI, depreciation charges, working capital and other variable costs should necessarily be identical for the OPDC and industrially developed countries because these firms would normally have different objectives and expectations from their projects and the returns on them, especially in the early stages of development. This is perhaps best expressed by the words of the Kuwaiti oil minister when addressing an OPEC meeting in 1978: 'investments in these countries [OPEC] should not be viewed on purely commercial considerations, they must also take into account national and social benefits.'[35] The OPDC would also be expected to reduce their high capital costs in the future as their infrastructures become more fully developed, removing the existing bottlenecks and as they improve their skills, gaining experience in operating and constructing petrochemical plants.

Nevertheless, some critics[36] argue that the projects being developed in the OPDC, especially the export-oriented ones in the Middle East, are being unduly subsidised by underpriced feedstocks and the soft loans to joint ventures at low interest rates. However, the gas feedstocks, around which most of the new plants are being built in the OPDC, are currently being wasted (flared). Also, it would be uneconomic to liquefy methane and ethane, whose net back value as fuels or LPG for export to distant places would be very low.

On the other hand, one could argue that industrialised countries directly and indirectly subsidise their industries, notably the steel and car industries in Western Europe, or offer government grants to industries in special 'development areas'. In the early 1980s, the British government allowed Shell, Exxon and BP to assume an appropriate transfer price for their ethane feedstock, supplied from fields in the North Sea to major petrochemical complexes in Scotland, as an inducement for these companies to go ahead with their projects.[37] Canada, also, has a policy of differential gas pricing, where LNG for export is priced higher than that offered to the home petrochemical industry.

Although the export-oriented petrochemical capacities, planned to come on stream in the second half of the 1980s, are still very small relative to the existing world capacities, it is expected the OPDC will play a more significant role in the world petrochemical industry by 1990, while more than 35 per cent of the steam crackers in Japan and the EEC were 15 years old in 1985 (UNIDO, 1981) and need replacement. The increasing capital costs required for these investments mean that chemical firms would have to rely on external financial sources, but the low returns expected from these investments would dissuade these companies from increasing their involvement in bulk petrochemical production. Instead, they would direct their investments into more profitable downstream and speciality products (agrochemicals, pharmaceuticals, biochemicals, etc.), a process which is already under way.

Again, this situation would strengthen the position of oil companies and the OPDC in petrochemicals. The OPDCs role as exporters of petrochemical products would be enhanced, becoming perhaps the world residual suppliers of bulk petrochemicals in the same way as they are the world residual oil suppliers.

4.4. The main obstacles facing petrochemical exports from the OPDC

The production cost advantage enjoyed by the OPDC, accepting low rates of return on their investments (5 per cent ROI), does not necessarily guarantee them markets for their products in major consuming industrialised countries. The competitiveness of those products would depend on their landed cost-price advantage over local product prices in the markets of destination. This means that exports from the OPDC would have to bear additional costs of shipping and tariffs as well as overcoming other non-tariff barriers in their import markets and remain competitive with local production costs.

In the following section, the transportation costs and tariff charges on exports from two potential producing regions, exemplified by Mexico and Qatar, are examined to assess the possibility of future developments in exporting petrochemical products from oil-rich areas to major markets of industrialised countries.

4.4.1. Transportation costs

The transportation costs of petrochemical products depend on a combination of the following factors: the physical characteristics of the products, the type (bulk carrier, refrigerated, pressurised, etc.) and size of the carrier vessel required, the operating costs and charges (crew, maintenance, fuel, insurance, etc.) and, finally, the distance of the market of destination.

The freight costs (UNIDO and Drewry, 1982) for shipping various petrochemical products from Qatar and Mexico to Japan, Western Europe and the United States are presented in Table D4. The data show high costs for plastics polymers, especially when they are bagged, of about $90 to $120 per tonne, which represents just over 10 per cent of the production cost of these products. For urea, ethylene and ammonia, the freight cost is lower but the transportation cost represents between 20–40 per cent of their production cost. Ethylene oxide, styrene and ethylene glycol have significantly lower freight costs. In all cases, however, the shipping cost is strongly influenced by the use of varying or standard vessel sizes and the physical conditions (pressurised, refrigerated, etc.) under which the products are transported.

4.4.2. Tariff barriers

The second important obstacle facing exports from the OPDC to developed countries are tariff taxes, imposed by receiving countries on imported chemicals to adequately protect their national industries.

It was found that higher tariff rates were charged for polymers and end products than for basic products in Japan and the EEC, while in the United States basic petrochemicals were imported duty free. The values of these tariff rates are shown in Table D5.

4.4.3. The competitivity of petrochemical imports in major markets

When the extra costs of freight and tariff charges are added to the production costs (at 5 per cent ROI) of various products, originating from Qatar, it was found that the landed cost for plastics polymers (LDPE, HDPE), methanol, ethylene, ammonia, SBR and styrene would remain more competitive than the local production from similar newly-built plants in Japan, Western Europe and the United States, as shown in Table 4.4.

Similarly, Mexico would be able to export successfully a wide range of petrochemicals to Japan, Western Europe and the United States, including plastics materials, PVC, LDPE, HDPE, LLDPE, PP and PS, in addition to SBR and basic petrochemicals, as can be seen from Table 4.5.

It is also obvious that Middle East producers would be in a better position to export to Western Europe and Japan, while Mexico would be suited to export to Latin American countries and the United States, as dictated by the geographical position of these producers.

The imposition of high tariffs on imports of downstream petrochemical products from developing countries, in addition to freight costs, results on average in a further increase in prices of landed products of about 20–30 per cent, which has the effect of deterring the OPDC from exporting a wider range of intermediate and end products and depressing the profitability of those products in which they are able to compete.

There are also other obstacles facing exports of petrochemicals into markets of developed countries, which relate to the oligopolistic structure of these markets and the vertical integration of production processes (discussed in more detail in Chapter 2), making it difficult for OPDC producers to penetrate these markets. However, the involvement of some of the major chemical and oil-based petrochemical producers in many of the export-orientated projects in the OPDC would make it easier to accommodate these exports in downstream operations, of the joint-venture partner, in developed markets.

It is unlikely that the governments of industrialised countries would respond to calls from OPDC producers for more co-operation in speeding-up the process of restructuring the petrochemical industry to give a larger role to oil producers and to

Table 4.4 Production cost, shipping charges and tariffs influencing competitivity of petrochemicals from Qatar exported to industrialised country markets landed versus local costs (1980) (US$/tonne)

		Export market															
		Japan				Northern Europe				Southern Europe				USA			
Product	Production cost at 5% ROI	Shipping cost	Tariff	Total cost	Local cost at 25% ROI	Shipping cost	Tariff	Total cost	Local cost at 25% ROI	Shipping cost	Tariff	Total cost	Local cost at 25% ROI	Shipping cost	Tariff	Total cost	Local cost at 25% ROI
Ammonia	151		7	193	375	39	21	211	345	30	11	192	345	51	5	207	317
DMT	928		—														
Ethyl benzene	618		—														
Ethylene	290	43	20	353	746	48	21	359	918	36	20	346	918	62	0	352	612
Ethylene glycol	845		—														
Ethylene acid	695		—														
HDPE	737	91	91	919	1,479	95	104	936	1,379	78	102	917	1,379	119	92	948	1,061
LDPE	638	92	80	810	1,367	96	92	826	1,295	79	90	807	1,295	120	80	838	979
LLDPE	710	92	88	890	1,144	96	100	902	1,243	79	99	888	1,243	120	89	919	751
Methanol	111	19	6	136	352	22	17	150	313	17	17	145	313	29	20	160	281
PET	1,875		—	—	—	—	—	—	—	—	—	—	—	—	—	—	—
Polypropylene	—	92			1,283	96			1,129	79			1,129	120	181		986
Polystyrene	1,051	92	160	1,303	1,474	96	143	1,290	1,262	79	141	1,271	1,262				
PVC	1,343	85	86	1,514	1,473	88	179	1,610	1,810	73	177	1,593	1,310	120	131	1,302	1,068
SBR	1,671	92	0	1,763	2,334	96	53	1,820	2,286	79	52	1,802	2,286	111	136	1,590	1,090
Styrene	831		—											120	0	1,791	2,079
TPA	972																
Urea	134	39	—	—	—	41	—	—	—	33	—	—	—	54	—	—	—
VCM	639		—	—	—		—	—	—		—	—	—		—	—	—

Table 4.5 Production cost, shipping charges and tariffs influencing competitivity of petrochemicals from Mexico exported to industrialised country markets landed versus local costs (1980) (US$/tonne)

		Export market															
		Japan				*Northern Europe*				*Southern Europe*				*USA*			
Product	*Production cost at 5% ROI*	*Shipping cost*	*Tariff*	*Total cost*	*Local cost at 25% ROI*	*Shipping cost*	*Tariff*	*Total cost*	*Local cost at 25% ROI*	*Shipping cost*	*Tariff*	*Total cost*	*Local cost at 25% ROI*	*Shipping cost*	*Tariff*	*Total cost*	*Local cost at 25% ROI*
Ammonia	126	51	7	184	375	29	17	172	345	48	19	193	345	8	4	138	317
DMT			—	—	—		—	—	—		—	—	—		—	—	—
Ethyl benzene	431		—	—	—		—	—	—		—	—	—		—	—	—
Ethylene	707	62	21	365	746	35	20	337	9.3	59	21	362	918	9	0	291	612
Ethylene glycol	707		—	—	—		—	—	—		—	—	—		—	—	—
Ethylene acid	581		—	—	—		—	—	—		—	—	—		—	—	—
HDPE	625	118	82	825	1,479	77	88	790	1,379	90	89	804	1,379	36	75	736	1,061
LDPE	539	119	72	730	1,367	78	77	693	1,295	91	79	709	1,295	37	67	643	979
LLDPE	606	119	80	805	1,144	78	85	770	1,243	91	87	784	1,243	37	75	719	751
Methanol	93	28	6	127	352	16	14	123	313	17	14	124	313	4	16	114	188
PET	1,615	—	—	—	—		—	—	—		—	—	—		—	—	—
Polypropylene	519	119	140	778	1,283	78	75	672	1,129	91	76	686	1,129	37	65	621	986
Polystyrene	774	119	125	1,018	1,474	78	106	958	1,262	91	108	973	958	37	132	942	1,068
PVC	796	110	54	960	1,473	72	108	976	1,310	84	110	990	1,310	34	80	910	1,090
SBR	1,255	119	0	1,374	2,334	78	40	1,373	2,286	91	40	1,386	2,286	37	0	1,292	2,079
Styrene	603		—	—	—		—	—	—		—	—	—		—	—	—
TPA	876		—	—	—		—	—	—		—	—	—		—	—	—
Urea	109	54	—	—	—	32	—	—	—	35	—	—	—	12	—	—	—
VCM	646		—	—	—		—	—	—		—	—	—		—	—	—

Source: UNIDO 1981.

remove or ease tariff barriers on some of the petrochemical products. It is possible that their commitment to minimum government intervention would lead them to leave the industry to reshape itself according to market forces. To become more competitive in their export markets OPDC petrochemical producers, therefore, would have to improve their efficiency in operating their plants and reduce their high capital costs by developing their own skills and ironing out their current deficiencies. Other oil producers, such as Saudi Arabia, could promote their petrochemical exports by relying on their strong bargaining power as major oil exporters, which is, perhaps, what the SABIC vice chairman had in mind when addressing the 1983 Plastics and Rubber Institute's conference (Erlichman, 1978): 'If protectionist measures of price manipulations are used to keep us out of the [petrochemical] market, then we shall counter with our own measures to obtain our rightful position in the market.' However, attempts by Solvay and ATO-Chimie to form a European petrochemical cartel to face the industry's problems and the expected increase in competition from OPDC producers were rejected by other major European chemical producers.[38] However, governments of the developed countries, especially those of the EEC, would still impose a variety of additional barriers on certain imports of petrochemicals, such as import quota and various forms of taxes and duties to protect their home industries. This year the EEC imposed, under the provisions of the Generalised Scheme of Preferences (GSP), a 13.4 per cent common customs tariff on imports of LLDPE from Saudi Arabia*. Earlier shipments of methanol from Saudi Arabia and Libya were subjected to a similar tax.

The full decontrol of oil and gas prices, coming into effect by the mid-1980s, would, perhaps, make American chemical users more willing to import large quantities of energy-intensive petrochemicals from more competitive producers such as Canada, Mexico and Saudi Arabia, where American oil companies are involved in joint ventures.

On the other hand, Western European producers have current and planned capacities that come on stream by the late 1980s, which would be sufficient to meet most of their petrochemical demands until 1990. However, the major producers are becoming

* *Source*: *European Chemical News*, 12 August 1985, p. 4.

more aware of the developments in the OPDC and have been closing down old plant and reshaping their industries accordingly to accommodate for the loss of some of their export markets in the future and the imports from the OPDC that will find their way into the European market. According to Robert Horton, the chairman of BP chemicals, about 20 per cent of the total plastics on the European market could originate from plants in the Middle East before the end of this century (Horton, 1981).

It is difficult to forecast exactly future developments in the petrochemical industry or to determine the level of penetration of import markets by OPDC producers, since the process of restructuring the petrochemical industry is still in its early stages. However, it is certain that developing countries would continue to increase their involvement in petrochemical production and to rely on local production to meet their petrochemical needs, which is evident from the large number of plants planned to come on stream in many of these countries. This is the subject of the following section of this chapter.

4.5. The supply of and demand for major petrochemicals in developing countries

4.5.1. Basic petrochemicals

The production of petrochemical products, namely basic petrochemicals and plastics materials, in developing countries was until the early 1970s very small. The production capacities were also small by world standards and were concentrated mainly in industrialising countries of Latin America and East Asia. The capacities of developing countries to produce basic petrochemicals in 1975 were about one million tonnes each for ethylene, benzene and propylene, while for butadiene, xylene and methanol they were less than 0.3 million tonnes, corresponding to between 3–5 per cent of total world capacities for these products.

Capacity has been increasing rapidly throughout the 1970s as a result of the changing structure of production costs of basic petrochemicals, brought about by oil price increases and the internationalisation of the industry as the process technology of these products was becoming mature and increasingly available to developing countries under licence from oil companies or chemical engineering firms. Existing and planned capacities, for ethylene

Table 4.6 Share of developing countries in total world production of selected petrochemical products

Petrochemical product	*World production* (million tonnes)*				*Developing countries output (million tonnes)*				*Developing countries share (per cent)*			
	1975	*1979*	*1985*	*1990*	*1975*	*1979*	*1985*	*1990*	*1975*	*1979*	*1985*	*1990*
Basic petrochemicals												
Ethylene	24.4	37.6	38.0	45.7	1.15	2.73	3.75	7.50	4.7	7.2	9.8	16.4
Propylene	12.6	19.7	21.4	25.5	0.47	1.19	1.75	2.40	3.7	6.0	8.2	9.4
Butadiene	3.4	5.0	4.9	6.3	0.20	0.40	0.50	0.70	5.8	7.9	10.2	11.1
Benzene	11.3	17.2	17.3	19.5	0.68	1.18	1.43	1.95	6.0	6.9	8.2	10.0
Xylenes	3.8	6.1	6.1	7.3	0.16	0.66	0.60	0.80	4.2	10.8	9.8	11.0
Methanol	7.5	11.6	13.5	16.7	0.25	1.20	2.80	4.80	3.3	10.3	16.7	28.7
Thermoplastics												
LDPE	7.5	12.2	13.4	17.2	0.5	1.1	2.10	3.70	6.1	8.9	15.7	21.5
HDPE	3.2	5.8	7.2	9.1	0.1	0.3	0.80	1.50	2.6	5.0	11.1	16.5
PP	2.3	5.0	6.7	9.3	0.05	0.3	0.60	1.20	2.1	7.1	8.9	12.9
PVC	7.6	12.2	13.4	16.5	0.7	1.6	1.70	3.20	8.8	13.1	12.7	19.4
PS	3.8	5.9	6.3	7.5	0.2	0.4	0.65	1.10	5.0	7.1	10.3	14.6
Total	24.4	41.1	47.0	59.6	1.45	3.7	5.85	10.70	6.0	9.1	12.4	17.9

* Including E. Europe and USSR.

Sources: UNIDO 1981, p. 38; UNIDO, IS. 427, 1983; SRI.

Table 4.7 Estimated annual rate of growth of demand for basic petrochemicals in developing countries (rate per annum)

	1975–9	*1980–85*	*1985–90*
Ethylene	24.0	7.5	7.4
Propylene	26.0	7.0	6.3
Butadiene	18.9	7.5	7.8
Benzene	12.8	7.2	5.9
Xylene	41.4	9.5	8.5

Source: Table 1.5.

and methanol, for example, could, on the basis of existing plans, reach 10.5 million and 5.3 million tonnes respectively (Table 4.6 and Table D6).

The share of each developing country in the production of basic petrochemicals is given in more detail in Table D6, which is arranged on a regional basis. The table shows clearly that the oil-producing countries and the NICs have a dominant role to play in petrochemical production.

The pattern of growth in the production of these products is shown in Table 4.7 and is similar to that in industrialised countries, where very high growth rates are achieved in the early stages of the industry's development which slow down later on, but continue to grow at relatively high rates until the markets become saturated.

Demand for basic petrochemicals in developing countries was rising at very high rates in the late 1970s. Growth rates of 7–9.5 per cent per annum are expected in the 1980s, if the demand for major plastics is to be met by local production. Ethylene production could rise from less than 3 million tonnes in 1979 to about 7.5 million tonnes in 1990. Similarly, the production of propylene, benzene and methanol could rise from about one million tonnes to between 2 and 4.5 million tonnes each, as shown in Table 4.6. These expansions would correspond to a rising share for the developing countries in the production of basic petrochemicals from about 7 per cent to 10–11 per cent of the total world output over the same period, as shown in the table; for the two major products, ethylene and methanol, the share would be higher at 16 and 28 per cent respectively.

Table 4.8 Demand for major plastics in developing countries (million tonnes)

Region	*1975*	*1980*	*1985*	*1990*
Asia*	1.20	2.60	3.75	5.13
Latin America	1.10	1.94	2.81	3.53
Middle East and N. Africa	0.44	0.89	1.31	1.85
Other	0.10	0.16	0.33	0.48
Total	2.84	5.59	8.20	10.99

* China and centrally planned Asian nations not included.
Sources: UNIDO 1981, p. 51; UNIDO, IS. 427, 1983; SRI.

4.5.2. Major plastics materials

The production of the five major plastics products (LDPE, PVC, HDPE, PP, PS) until the mid-1970s, with the exception of a very small PVC and HDPE production in Algeria, Venezuela and Egypt, has been confined to the NICs, which will continue to be the major producers among the developing countries until the end of the 1980s. However, many developing countries are expected to have production capacities coming on stream in the second half of the 1980s, particularly for LDPE and PVC. The largest contributions will be made by the countries with rising industrial outputs (Brazil, Mexico, Argentina, South Korea and India) and the oil producers of the Middle East. The contributions made by these developing countries are given in more detail in Table D7.

The demand for major plastics products in developing countries is estimated to increase by 5.5–8 per cent per annum, increasing from about 6 million tonnes in 1980 to about 11 million tonnes in 1990. The distribution of this demand on a volume basis among the various regions is shown in Table 4.8.

The products most widely in use in developing countries are LDPE and PVC, which account for more than 60 per cent of total demand for plastics materials. Their production in developing countries is expected to exceed 3 million tonnes each in 1990. While PP, which is growing at a higher than average growth rate, could reach 1.2 million tonnes.

With the rising demands for petrochemicals and more capacities coming on stream, the developing countries could become self-sufficient in the production of major plastics materials by 1990, when they could account for between 13–20 per cent of the total world production of these products (see Table 4.6). However, developing countries would continue to import large quantities of intermediate and end petrochemical products such as solvents, fibres, speciality polymers and plastics, etc., from major producing companies of industrialised countries. Trade in the major basic petrochemicals and plastics materials would be increasing during this period between the various groups of countries, including the major producers, to meet the imbalances that may occur in their supply and demand pattern.

4.6. Concluding remarks

The 1970s saw the unfolding of a new chapter in the development of the petrochemical industry with expanding petrochemical production in developing countries in a process of change brought about by successive oil price rises which altered the production cost structure so that raw materials costs and the availability of the process technology from imitator-innovators such as the major oil companies and the chemical engineering and construction firms were much more significant.

Two groups of developing countries were very active in petrochemical production: the major oil producers and the newly industrialised countries. Among the oil producers, the development of the petrochemical industry was proceeding at two levels. First, the sparsely populated but oil-rich countries of the Middle East were seeking to become major export-oriented producers of petrochemicals, utilising the vast quantities of associated gas which until recently had been largely flared. To achieve their objectives, these countries have been offering generous terms to major oil and chemical companies to encourage them to undertake substantial investments, on a joint-venture basis, in the massive petrochemical complexes that are under construction or are at the planning stage in these countries. Second, the oil-producing countries, which have larger home markets, an already existing infrastructure and a not inadequate supply of indigenous skilled labour operating in the oil sector, have been taking a more modest approach in the development of their petrochemical industry, building medium-sized plants whose production is aimed at meeting the demands of the home market and the expanding industries, with the excess production aimed at regional or international export markets.

The NICs, notably Brazil and South Korea, were able to develop their petrochemical industry at an earlier stage and at a faster rate than oil producers and other developing countries, aided by the growing demand for petrochemicals from their larger home markets and their rapidly expanding industrial outputs.

Japan and the major oil companies, realising the changing situation in feedstock costs and availability, have been active in participating in petrochemical projects in oil-producing countries, such as Iran and Saudi Arabia as well as in major oil-processing centres such as Singapore, to ensure the supply of basic petrochemicals for their home down-stream end-processing industry. This would also ensure those producers of markets for their end products in countries in which they are involved in joint ventures.

In 1982 Dow pulled out of a major joint-venture commitment for LDPE production in Saudi Arabia. This was not due to uneconomic viability or to the terms under which the joint venture was concluded, becoming unattractive. Dow was at the time also pulling out of major projects in bulk petrochemicals in Yugoslavia and Canada in an effort to reduce its debt load, which was standing at 51 per cent, and to move into more profitable downstream operations.[39]

The major oil producers of the Middle East, Mexico and Indonesia, realising the comparative advantage they now possess in the production of basic petrochemicals and various plastics materials, whose production costs are dominated by the feedstock component, are aiming at becoming major world producers of petrochemicals, with a substantial proportion of their outputs aimed at the export markets. To be successful in their efforts, these countries should ensure that they reduce their present high capital costs and overcome the high charges of freight and tariffs on products destined for major consuming markets. This can be achieved by using large-size chemical carriers that could also be adapted to carry back goods on the return trip, thus significantly reducing freight charges; while tariffs could be reviewed under the terms of bilateral or regional agreements.

It must be emphasised, however, that it is premature at this stage to forecast with any accuracy the impact that these developments will have on the global structure of the petrochemical industry and the respective roles various operators will play, since the process of restructuring the petrochemical industry is still in its early stages. However, it is widely accepted that oil-rich countries, particularly those of the Middle East, will increase their share in world petrochemical output and export markets, especially in basic petrochemicals and bulk polymers, in which they have a clear comparative advantage despite their present high capital costs. Should energy prices rise in the 1990s decade, it would again squeeze the profits of major chemical producers, forcing them to go further down-stream where higher profits can be achieved leaving the arena of bulk petrochemical production open to OPDC producers and major oil companies.

5 CONCLUSIONS

This book has been concerned with the study of the development stages of the petrochemical industry within the framework of technological and economic factors that led to the reshaping of the industry to its present form. Despite the fact that the statistical data presented throughout have not been treated in a sufficiently analytical way, due to the lack of sufficiently consistent data, and in view of the variety of aspects covered and the brevity of scope of this study, every effort has been made to point out the developing trends and the changing structure of the industry, its production costs, and the roles played by the main actors involved in petrochemical production.

The modern petrochemical industry was developed around the oil-processing industry, both of which grew rapidly with the development of the fluid catalytic cracker's technology and the expansion of the oil-refining capacities after the Second World War. The clustering of innovations in the 1920s–40s and again during the 1950s laid the foundations for the production of a variety of new petrochemical products, notably the synthetic materials such as plastics, fibres and rubbers, whose production has increased rapidly from a few thousand tonnes in the 1950s to tens of millions of tonnes in the 1970s as a result of the rapid technological improvements and the rising demand and growing markets for these products, whose prices were more competitive and whose qualities were superior to those of conventional materials they were replacing. It has already been established that the major chemical companies were responsible for most of the innovations (Freeman, 1974), which led to the production of new or modified products and improved processes of production.

An innovating firm enjoys monopolistic profits in the early years of production of a new product, which allows it to reinforce its position in the market and assert its leadership. This situation encourages, or rather forces, other competing firms, which are usually equally important and leading chemical producers, to develop their own technologies for the production of the new products, since the innovating companies would be willing to license their new production processes only to other firms in which they have a majority share (Stobaugh, 1968). This has been the case with products such as nylon, PVC, HDPE, and, more recently, LLDPE.

In this way, a *swarm* of producers is formed and production capacities start to increase as imitator producers (which form the *swarm*) join the innovator producer in the production of new products. Throughout this period, these products and their processes of production undergo improvements and modifications, with larger plants being built all the time to benefit from the economies of scale, thus reducing their production costs. Eventually the process of production becomes standardised and the production units widespread (Section 1.4, Chapter 1).

Non-producing imitators, such as engineering firms, play an important role in this process since they have no production capacities of their own. They contribute to the process of internationalising petrochemical production by licensing to a large number of producers in various countries processes they developed themselves or obtained under licence from the innovator at a stage where the process had become mature and widespread.

It has been argued throughout this study that two main factors contributed to the increased maturity of the industry. The first is the natural evolution of the industry, resulting from the diffusion of the basic technologies by imitator producers and engineering firms, as described above. The second is the changing structure of production costs so that feedstock costs are more important, which results from the oil price increases of the 1970s. This reinforced the position of the imitator producers, mainly chemical subsidiaries of oil companies and government-backed companies in some developed markets (Italy and France), accelerated the 'maturation' of the industry and changed the roles of main operators and producers of petrochemical products.

The oil price rises of the 1970s were also accompanied by mounting uncertainty about the security of feedstock supplies to chemical majors, where these had to compete with the automotive industry for their supplies of the light oil fractions. This situation was more serious in Western Europe and Japan, where the industry has been largely based on naphtha as a feedstock, than in the United States, where natural gas has been the main feedstock.

These developments (increasing number of producers, maturity of the industry, rising oil prices and the uncertainty of feedstock supplies) led the major chemical companies to reconsider their role in the petrochemical production process. On the one hand, these companies were integrating backward into refining to secure their feedstock supplies for their existing and planned petrochemical plants. At the same time, they were taking appropriate measures to go further downstream in their petrochemical operations to produce speciality products, pharmaceuticals, biochemicals, agrochemicals, etc., where higher profits can be achieved, since the profit margins in basic petrochemicals and bulk plastics products have been squeezed by increased competition from imitator producers and the standardisation of the production processes of these products. The major chemical companies have realised that the role of the oil companies in future petrochemical production would become more significant, since the competitive advantage in the production of basic petrochemicals has moved in their favour due to their flexibility in obtaining the right mix of feedstocks from their integrated refinery-petrochemical complexes.

The oil companies, in turn, found an opportunity to diversify from oil production and increase their returns on hydrocarbons by increasing their involvement in petrochemical production, helped by their control over oil supplies and their increased cash flows, following the oil price rises, their ability to develop or obtain under licence the required process technology, and their marketing expertise (Sections 2.5 and 2.6, Chapter 2).

Capital investment costs for petrochemical plants continued to rise sharply throughout the 1970s which, together with the very large capacity plants that were coming on stream, meant that the financial requirements for such plants have become extremely large and laid the foundation for cooperation between chemical companies and oil majors by forming joint ventures for ethylene and other petrochemical production, thus spreading the financial risks and commitments. The chemical companies found in these joint ventures an opportunity to secure some of their basic petrochemical requirements and feedstocks, while for oil companies these ventures opened the way for increased petrochemical involvement with direct access to the markets.

Another group of operators, which became active in petrochemical production during this period, were government-backed companies in Italy and France, whose share of petrochemical output increased steadily because they had secure and subsidised feedstock supplies from state-owned oil companies.

The access to feedstocks has also been the main driving force behind joint ventures between oil or chemical majors, on one side, and petrochemical operators of oil-rich countries, on the other. The Japanese petrochemical companies, realising their vulnerable feedstock position, were very active in these joint ventures, particularly with oil-rich countries of the Middle East (Section 4.2.3, Chapter 4).

The maturity of the industry and the changing structure of production costs have altered the roles of major petrochemical producers and started a process of restructuring the operations of the petrochemical industry. In the short run, this process, together with the effects of the recession, presented the industry in major producing regions with a number of problems in the late 1970s and the early years of the 1980s. For example, overcapacity resulted from the imbalance between the supply and demand for petrochemicals and the industry responded by cutting production capacities to restore the supply-demand balance. While in the United States and Japan capacities in some basic petrochemicals were cut by 15–20 per cent, capacity reductions were, however, more widespread in Western Europe in the early 1980s, mainly in basic petrochemicals such as ethylene and bulk polymers such as PVC and LDPE. Overcapacity was mainly a Western European problem, where a large number of new plants came on stream in the second half of the 1970s, following the oil price rise. Oil companies had a significant share in these plants. Also, these new plants were very large in scale. However, after the low output levels of 1975 the early signs of demand recovery during 1976–9, which the industry mistook as a return to the heydays of the 1960s when petrochemical output was growing at an average annual rate of more than 15 per cent, were short-lived and the world plunged into a deep recession; industrial production stagnated or declined in many countries, leading to a sluggish growth in demand for petrochemical products.

In addition to the large increases in capacity and the slow growth of demand, many of the old plants, which are considered by the industry's standard economic terms to be obsolete, were still in operation, adding to the problem of overcapacity. It was mainly these plants that were affected by the capacity cuts during the 1980–2 period, when the industry was taking measures on an individual company level to reshape itself by closing down the excess capacity so as to operate the remaining plants at full capacity, thus reducing their operational losses and at the same time curbing the increased competition and boosting product prices.

The production and consumption patterns of the major petrochemical products have been analysed on the regional level for North America, Western Europe and Japan, and the world as a whole (Section 1.6, Chapter 1) as well as at the country level for the United Kingdom and other major Western European producers (Section 3.7, Chapter 3). The analysis shows that the consumption or demand growth rates have been declining from the previously high levels of the 1960s as the markets for these products were becoming saturated. Demand for basic petrochemicals and plastics materials is, however, expected to grow at modest rates during the 1980s, as conventional materials such as steel, copper, glass, paper, etc., continue to be replaced by plastics and other synthetic materials, which save weight and consume less energy per item of product than the conventional materials which they replace (see Section 1.7, Chapter 1). There are also potentially large markets for plastics materials in the car industry and in the building industry, as the drive for weight- and energy-saving continues to gain momentum (Section 1.8, Chapter 1).

In forecasting future demands for petrochemical products, it has been found that too much reliance on the general indicator of GDP leads to inaccurate results, and instead, demands for petrochemical products are found to depend on the performance of the industrial manufacturing output, namely motor vehicles, electrical products and components, chemicals as well as packaging and the building industry. This has been demonstrated with a brief analysis of the United Kingdom plastics materials industry, where, since 1979 the expanding oil sector has been responsible for much of the GDP growth, while the output of the manufacturing industry,[40] plastics materials and petrochemical products has been declining. A forecasting procedure, employing an input–output model (Section 3.5, Chapter 3) would be more appropriate in this case since the output of the petrochemical industry is used mainly as intermediate inputs to other industries. The I/O type of analysis allows the effects of final demand components and linkage effects between various industrial sectors, comprising the economy, to be taken into account in relation to the output of the industry under investigation (Sections 3.2–3.4, Chapter 3). Input–output models also have the advantage of allowing modifications or updating procedures to be introduced into the model as well as the application of alternative scenarios in line with the economic outlook and the changing structure of industrial production.

Chapter 4 showed how the changing structure of production as feedstock became more expensive and the standardisation of the process technology and its increased availability from imitator producers (chemical companies and oil majors) and non-producing imitators (engineering firms, see Figure 4.1, Chapter 4) has helped to internationalise petrochemical production. Since the mid-1970s, an increasing number of developing as well as some of the smaller industrialised countries have been installing petrochemical plants for the first time or expanding their existing capacities, where demands for their markets can now be supplied from their small- or medium-size plants which they can operate at full capacity. In developing countries, production of petrochemicals and demand for their products would continue to grow at relatively high rates throughout the 1980s, since the production would be growing from a low base and there would be available large markets to be satisfied.

More important have been the implications of the changing structure of production costs for oil-producing countries, where these have access to cheap oil supplies and vast quantities of associated gas, which until recently was being flared at incredible rates. These countries, despite their currently high capital investment costs, can produce a large number of basic petrochemicals and some bulk plastics materials more cheaply than major producers of industrialised countries (Sections 4.3 and 4.4, Chapter 4), and if energy costs continue to increase in the future, then the oil-producing developing countries will have the opportunity to play a more significant role in the production and export of petrochemical products on the global scene during the 1990s.

However, it should be emphasised that the existing oligopolistic structure of the petrochemical industry and its markets (Section 2.7, Chapter 2, and Section 3.8, Chapter 3) will probably hinder any major or swift changes in the roles of petrochemical producers. Nevertheless, the process of restructuring petrochemical production, where oil companies and oil-producing countries would become major petrochemical producers, is continuing slowly but steadily, while major chemical companies move into more profitable areas of production with better prospects of growth.

The findings of this study confirm that market size and industrial output are the main determinants for the development of a petrochemical sector. Brazil and South Korea have no hydrocarbon resources of their own, yet their petrochemical industry was developed at an earlier stage and their petrochemical output at present exceeds that of any major oil-producing developing country (Section 4.2, Chapter 4). This has been mainly a result of the large market size and the rising industrial output of these two countries which, in turn, resulted from their concerted efforts to industrialise and the specific policies adopted by them to create an indigenous scientific and industrial base, supported by appropriate institutional and infrastructural arrangements.

Finally, the development of a petrochemical industry in oil-producing developing countries should be undertaken as a part of their industrialisation process as a whole to ensure its success and compatibility with the development plans of these countries, avoiding severe economic bottlenecks and overstretching their human and physical resources.

Oil-producing developing countries, which have chosen to develop their petrochemical industry by joint ventures with some of the world-leading oil or major chemical companies, should pay particular attention to the development of their own managerial as well as technical skills so that benefits from these ventures contribute equally to the producing country and the joint-venture partner, since too much reliance on a multinational company restricts the options that are open to the OPDC in a variety of ways. For example, it can limit their export markets, production outputs of certain products and expansion of capacity, which are usually classified in the licensing agreements. Also, a prolonged reliance on a large and unsettled expatriate work-force does not make much economic sense in the long run.

There are challenging and yet not easy tasks facing the OPDC, especially those with ambitious industrialisation plans, for example, Middle East countries and Mexico. On the national level these require careful planning, a sincere and able management, a capable and skilful working force and the determination to succeed; and, on the global level, increased regional and international co-operation as well as a healthy economic environment. The achievements and the degree of success of these countries in meeting the targets of their plans have yet to be seen.

REFERENCES AND NOTES

1. Industrial sources have explained that the justification for this process of building new plants, at a time when demand growth was slowing down, is that the development plans for these plants had been planned over a long period and that by the time they were due to come into effect it was too late to change the plans or scrap them completely.
2. OECD, 1979. See Figures 4 to 6 in the statistical annex.
3. The UNIDO Study (UNIDO IS. 427, 1983), which has been developed for estimating the world demand for petrochemical products, employs an econometric model of the basic structure of the petrochemical industry. Its demand equations are of the derived demand form, where estimates for basic petrochemicals are derived from projections of demand for intermediate and final products.
4. See *Petroleum Times*, November 1981, p. 22.
5. These four products account for about 70 per cent of the total production of plastics materials.
6. For a wider treatment of this subject see article by Dale Rudd (1975).
7. *Plastics Today*, No. 7, Winter 1979–80, p. 19.
8. Ibid., p. 21.
9. *Plastics Today*, 'Plastics after the Oil Crisis: the ICI View', Summer 1974, p. 14.
10. IEA, 'A Voice Against Complacency', *Financial Times*, 13 October 1982, p. 12.
11. J. Cadogan, 'Chemistry in the Petroleum Industry', BP Symposium at Imperial College, 9 March 1983.
12. *Modern Plastics*, January 1980, p. 103.
13. *Plastics Today*, 'Plastics after the Oil Crisis: the ICI View', December 1978, p. 14.
14. *Hydrocarbon Processing*, February 1981, p. 45.
15. *European Chemical News*, 'Editorial Comment', January 1980.
16. For capital investment data on the oil industry for the largest twenty-six oil companies in the world, including the seven sisters, and the national oil companies, see 'The Petroleum Situation', Chase Manhattan Bank, January 1982.
17. *Oil and Gas Journal*, 'Ethylene Report', September 1980.
18. UNIDO, 'First world-wide study on the petrochemical industry: 1975–2000', ICIS. 83, December 1978, p. 77.

19. *European Plastic News*, 3 October 1975.
20. *Plastics Industry Europe*, Vol. 7, No. 5, March 1983.
21. The industrial sectors studied in this chapter are referred to by their Standard Industrial Classification (SIC) Numbers, based on the revised 1968 edition. Business Monitor, *Input–Output Tables for the United Kingdom 1974*, PA1004, HMSO, 1981, p. 34.
22. Ibid., p. 42.
23. Table D is an industry × industry table which shows how each industry's output depends on direct inputs from supplying industries. The sum of the elements row (29) form the total output of the synthetic resins and plastics materials industry going to other industries and final demand components. These elements in turn represent the plastics materials inputs to the industries of the columns in which they fall. See Business Monitor, *Input–Output Tables for the United Kingdom 1974*, ibid., p. 34.
24. Ibid.
25. The work carried out by the Group on input–output models usually appears in the 'Economics Reprints' of the Department of Applied Economics, University of Cambridge. See *Economic Reprints*, Nos. 7, 10, 11, 15 and 22.
26. Based on data obtained from *United Nations Yearbook of Industrial Statistics – 1980*, UN, New York, 1982.
27. *Oil and Gas Journal*, 'Ethylene Report', 7 September 1981, p. 85.
28. BP took over thermosetting capacity of BXL (a subsidiary of Union Carbide) in 1977. Also Monsanto's polystyrene capacity was transferred to BP early in 1979. *European Plastics News*, January 1979, p. 18. Similar developments were later reported in the United Kingdom and the Continent.
29. BP, private source.
30. Shell was reported as closing down its old petrochemical plants, *Financial Times*, 4 June 1982, p. 12. Also on 18 June 1982, p. 18.
31. *European Plastics News*, 'The UK Plastics Industry', January 1979, p. 15. Also, ICI and BP announced price rises in PVC and LDPE respectively, following their joint deals in those two products, reported in *Plastics Industry Europe*, Vol. 7, Nos. 5 and 9, 1983.
32. Based on data obtained from *Economic Trends*, No. 353, March 1983, p. 33, for the UK (1957) and from the *UN Yearbook of Industrial Statistics –1980*, for the United Kingdom and Western Europe, op. cit., p. 669.
33. *Hydrocarbon Processing*, February 1980, p. 62.
34. SRI International.
35. *The Economist*, 6–13 April 1979, p. 79.
36. Those critics include a wide spectrum of observers, those involved in petrochemical production as well as economists and industrial journalists. However, this subject is dealt with in a more serious way

by Louis Turner in *Middle East Industrialisation: A Study of Saudi and Iranian Downstream Investments,* Royal Institute of International Affairs, London, 1979.

37. The British Government's plan to allow Shell and BP to use transfer pricing for their ethane feedstock from the North Sea fields caused a big row between ICI, whose ethylene plants are naphtha based, and the Government, as the former saw this regulation as unfair subsidy. This was reported widely by the media during the summer of 1982. Later on, however, the Government went ahead with its regulation.
38. *Financial Times,* 'Plastics Crisis Cartel', 23 July 1982, p. 2. Also *The Economist,* 24 July 1982, p. 58.
39. G. Gray, Petrochemicals Planning Department, BP Chemicals, London, 1983; *The Economist,* 4 November 1982, p. 78.
40. In 1982 the output of the manufacturing industry in the United Kingdom had declined to 88 per cent of that of 1975, *Economic Trends,* No. 353, March 1983, p. 26.

BIBLIOGRAPHY

Al-Wattari, A., *Oil Downstream: Opportunities, Limitations, Policies, Organisation of Arab Petroleum Exporting Countries*, Kuwait, 1980.

Anderson, E.V., 'The Arabs and their oil: will they make it into petrochemicals?', *Chemical and Engineering News*, 16 November 1970, pp. 58–73.

Asam, E.H., 'Forecasting the growth of industrial markets with Input–Output techniques', in Gielnik, S.J. and Gossling, W.F. (eds), *Input–Output and Marketing*, Input–Output Publishing Co., London, 1980.

Barker, T.S., *Some Experiments in Projecting Intermediate Demand*, Economics Reprint No. 10, Department of Applied Economics, University of Cambridge, 1977.

——— , *An Analysis of the Updated 1963 Input–Output Transactions Table*, Economics Reprint No. 11, Department of Applied Economics, University of Cambridge, 1977.

Barna, T., *International Conference on Input–Output Techniques: Structural Interdependence and Economic Development*.

Barnett, H., 'Specific industry output projections: long range economic projection', in Ghosh, A., *Experiments with Input–Output Models*, Cambridge University Press, Cambridge, 1966.

Barry, S. *et al.*, 'An international comparison of polymers and their alternatives', in Thomas, J.A.G. (ed .), *Energy Analysis*, IPC, Science and Technology Press, London, 1977.

Bennett, M.J., 'The future outlook for petrochemicals in Western Europe', for Chemical Systems International, in *The Petrochemical Industry Briefing*, Management Centre Europe, Brussels, 27–28 April 1982.

Bretsher, J.B., 'A marketing review', in Gielnik, S.J. and Gossling, W.F. (eds), *Input–Output and Marketing*, Input–Output Publishing Co., London, 1980.

Bucknall, C.B., *Toughened Plastics*, Applied Science Publishers, London, 1977.

Bulmer-Thomas, V., *Input–Output Analysis in Developing Countries: Sources, Methods and Applications*, Wiley, Chichester, 1982.

Business Monitor, *Input–Output Tables for the United Kingdom*, PA 1004, 1974, HMSO, London, 1981.

Cameron, B., *Input–Output Analysis and Resource Allocation*, Cambridge University Press, Cambridge, 1968.

Carter, A. and Brody, A. (eds), *Applications of Input–Output Analysis*, North-Holland, Amsterdam, 1970.

Caudle, P., 'Energy prices to the processing industries', *The Chemical Engineer*, February 1983, pp. 10–12.

CHEMFACTS, *United Kingdom, Chemical Data Services*, IPC Industrial Press Ltd., London, 1977.

Chase Manhattan Bank, *Capital Investments of the World Petroleum Industry: 1970–1980*, Energy Economics Division, Chase Manhattan Bank, New York, 1980.

______ , '1980 financial analysis of a group of petroleum companies', *The Petroleum Situation*, Vol. 6, No. 1, Energy Economics Division, New York, January 1982.

______ , *The Petroleum Situation*, Vol. 6, No. 2, Energy Economics Division, New York, July 1982.

Claydon, D.A., 'Petrochemical response to the market of the future', in Fwest, T. and Sharp, D. (eds), *The Chemical Industry*, Ellis Horwood Ltd., London, 1982, pp. 180–93.

Cox, J., 'Action programme for reorganisation', A Briefing, in *The Petrochemical Industry*, Management Centre Europe, Brussels, 27–8 April 1982.

Christ, C. *et al.*, *Measurement in Economics*, Stanford University Press, Stanford, CA, 1963.

Cramer, J.S., *Empirical Econometrics*, North Holland, Amsterdam, 1969.

Dafter, R., 'IEA: a voice against complacency', *Financial Times*, 13 October 1982.

Dataquest survey, *The Industrial Chemical Industry*, Jordan Dataquest Ltd., London, 1977.

Douglas, E., *Economics of Marketing*, Harper and Row, New York, 1975.

Economic Trends, No. 353, March 1983, HMSO, London, 1983.

Erlichman, J., 'Saudi's petrochemical assault', *The Guardian*, 13 June 1983.

European Plastics News, 'Bulk thermoplastics: long term strategies still needed', April 1984.

______ , 'Polystyrene: moving out of the Doldrums – but rationalization still needed', April 1984.

Fathi-Afshar, S., 'A systems study of the interchangeable petrochemical products', PhD dissertation, University of Wisconsin–Madison, 1979.

Fathi-Afshar, S. and Rippin, D.W.T., 'Game theory applied to chemical plant investment decisions in a competitive market', *Chemical Engineering Communication*, Vol. 26, 1984.

Fathi-Afshar, S. and Yang, J.C., 'Designing the optimal structure of the petrochemical industry for minimum cost and least gross toxicity of chemical production', *Chemical Engineering Science*, Vol. 40, No. 5, 1985.

Fischer, A., 'Engineering manpower costs more for smaller projects', *Hydrocarbon Processing*, April 1985, pp. 133–4.

Flavin, C., *The Future of Synthetic Materials, the Petroleum Connection*, paper No. 36, Worldwatch Institute, Washington D.C., 1980.

Freeman, C., *The Economics of Industrial Innovation*, Penguin, Harmondsworth, 1974.

______ , *et al.*, *Unemployment and Technical Innovation: A Study of Long Waves and Economic Development*, Frances Pinter, London, 1982.

Friedman, M., *Price Theory*, Aldine Publishing Co., Chicago, 1976.

Ghosh, A., *Experiments with Input–Output Models: An Application to the Economy of the United Kingdom, 1948–1955*, Cambridge University Press, Cambridge, 1964.

Griffin, J.M. and Teece, D.J., *OPEC Behaviour and World Oil Prices*, George Allen & Unwin, London, 1982.

Hahn, A.V.G. *et al.*, *The Petrochemical Industry: Market and Economics*, McGraw-Hill, New York, 1970.

Hatch, L.F., *From Hydrocarbons to Petrochemicals*, Gulf Co., London, 1981.

Heckle, M., 'European petrochemicals – overcapacity, low growth', *Petroleum Times*, November 1981, pp. 40–5.

Henderson, J.M. and Quandt, R.E., *Micro-Economics Theory: A Mathematical Approach*, 2nd ed., McGraw-Hill, New York, 1971.

Horton, R., 'Rationalisation long overdue in European petrochemical industry', *Petroleum Times*, November 1981, pp. 32–40.

Housing and Construction Statistics: 1971–1981, HMSO, London, 1982.

Howell, I.V., *Developments in Ethylene and its Major Derivatives*, MSc. dissertation, Imperial College, University of London, 1976.

Hufbauer, G.C., *Synthetic Materials and the Theory of International Trade*, Duckworth, 1966.

Hunter, D., 'Cheaper feedstocks anticipated by Europe's petrochemical maker', *Chemical Engineering*, 18 February 1985.

Hydrocarbon Processing, 'Outlook for the chemical industry', April 1985.

Jimenez, A. and Rudd, D.F., 'Use of a recursive mixed-integer programming model to detect an optimal integration sequence for the Mexican petrochemical industry', paper submitted to *Computers and Chemical Engineering*, 1984.

Kemeny, J.G., *Finite Mathematics with Business Applications*, Prentice Hall, Englewood Cliffs, N.J., 1962.

Koopmans, T. (ed.), *Activity Analysis of Production and Allocation: Proceedings of a Conference*, Cowles Monograph No. 13, Wiley, 1951.

Lanz, K., *Around the World with Chemistry*, translated from German by David Goodman, McGraw Hill, 1980.

Leontief, W., *The Structure of the American Economy, 1919–1939: An Empirical Application of Equilibrium Analysis*, Oxford University Press, London, 1951.

Liebeskind, D., *Forecasting Industrial Chemical Prices Within a Planning Framework Using Disaggregation of Input–Output Tables*, PhD dissertation, New York University, 1972.

Livesey, D.A., *A Minimal Realization of the Leontief Dynamic Input–Output Model*, Economics Reprint No. 7, Department of Applied Economics, University of Cambridge, 1976.

Mabro, R. (ed.), *World Energy Issues and Policies: Proceedings of the First Oxford Energy Seminar*, Oxford University Press, Oxford, 1980.

Mariotti, S., 'A nonlinear programming model to evaluate multi-product multi-plant efficiency of an industry', *Engineering Costs and Production Economics*, Vol. 7, 1984.

Mathis, J.F. and Brownstein, A.M., *CEP*, December, 1984.

Motamen, H., *Macroeconomics of North Sea Oil in the United Kingdom*, Heinemann, London, 1983.

National Economic Development Office (NEDO), *Investment in the Chemical Industry*, a report by the Investment Working Party of the Chemicals EDC, NEDO, London, 1972.

______ , *The Plastics Industry and its Prospects*, a report of the Plastics Working Party of the Chemicals EDC, NEDO, London, 1972.

______ , *Chemicals: Industrial Review to 1977*, NEDO, London, 1973.

______ , *UK Chemicals 1975–1985: Strategies and Opportunities for the Industry*, NEDO, London, 1976.

______ , *Progress Report to the National Economic Council*, Petrochemicals Sector Working Party, July 1981, unpublished.

Nerlove, M. *et al.*, *Analysis of Economic Time Series: A Synthesis*, Academic Press, New York/London, 1979.

OECD, *The Petrochemical Industry: Trends in Production and Investments to 1985*, OECD, Paris, 1979.

Ogorkiewicz, R.M., *The Engineering Properties of Plastics*, ICI Plastics Division, Wiley-Interscience, London, 1977.

OPEC, *OPEC and Future Energy Markets*, Proceedings of OPEC seminar held in Vienna, Austria, October 1979, Macmillan, London, 1980.

______ , *Annual Report – 1981*, OPEC, Vienna, 1982.

______ , *Annual Statistical Bulletin – 1982*, OPEC, Vienna, 1982.

Philpot, J.A., 'Technology, Costs and Prices', for Chemical Systems International, in *The Petrochemical Industry Briefing*, Management Centre Europe, Brussels, 27–8 April 1982.

Plastics and Rubber International, 'Linear Low Density Polyethylene: a state-of-the-art review', Vol. 10, No. 2, April 1985.

Quinland, M., 'In search of profits', *Petroleum Economist*, July 1982, p. 286.

Reader, W.J., *Imperial Chemical Industries: A History*, Oxford University Press, London, 1975.

Reining, F.E., 'Corporate financial considerations', in *The Petrochemical Industry Briefing*, Management Centre Europe, Brussels, 27–8 April 1982.

Rudd, D., 'Modelling the development of the intermediate chemicals industry', *The Chemical Engineering Journal*, No. 9, 1975, pp. 1–20.

Rudd, D. *et al.*, *Petrochemical Technology Assessment*, Wiley-Interscience, New York, 1981.

Sharp, D. and West, T.F. (eds), *The Chemical Industry*, Ellis Horwood, London, 1982.

Shearer, P., 'Using cyclical effects to improve business planning', *European Plastics News*, April 1984.

Shell International Petroleum Company, *Shell Briefing Service*, Shell Centre, London, various issues 1976–1980.

Short, H. and Chowdhury, J., 'Varied menu is a must for olefin makers', *Chemical Engineering*, 11 June 1984.

Smith, D., 'Downstream petroleum industries – the third phase', *Middle East Annual Review – 1977*, pp. 41–5.

Smith, R.P., *Consumer Demand for Cars in the U.S.*, occasional papers, No. 44, Department of Applied Economics, Cambridge University Press, 1975.

Stobaugh, R.B., *The Product Life Cycle, United States Exports and International Investment*, PhD dissertation, Harvard University, 1968.

______ , *Nine Investments Abroad and their Impact at Home: Case Studies on International Enterprises and the US Economy*, Harvard, 1976.

Stone, R., *Input–Output Relationships between a Main Model and a Submodel, Proposal for an I/O Analysis of the Chemical Industry in Europe for 1968, 1975, 1980*. A paper presented at the Battelle Institute, Geneva, undated.

Stork, K. *et al.*, 'Petrochemical – challenges for 2000', *The Oil and Gas Journal*, August 1977, pp. 442–50.

TECNON (UK) Ltd., *Western European Petrochemical Industry: 1981 Aromatics Report*, Parpinelli Tecnon, Milan, 1981.

Turner, L., 'Petrochemicals, refining and gas exports – problems with costs and marketing', *Middle East Annual Review – 1978*, pp. 87–90.

Turner, L. and Bedore, J., *Middle East Industrialisation: A Study of Saudi and Iranian Downstream Investments*, Royal Institute of International Affairs, Saxon House, Farnborough, 1979.

UNIDO, *First World-wide Study on the Petrochemical Industry: 1975–2000*, UNIDO, Vienna, UNIDO/ICIS. 83, December 1978.

______ , *First Consultation Meeting on the Petrochemical Industry*, UNIDO, Mexico, Report, March 1979.

______ , *First Draft of UNIDO Model Form of Agreement for the Licensing of Patents and Know-how in the Petrochemical Industry*, UNIDO, Turkey, ID/WG. 336/1, February 1981.

______ , *Long-term Arrangements for the Development of the Petrochemical Industry in Developing Countries Including Arrangements for Marketing Petrochemicals Produced in Developing Countries*, UNIDO, Turkey, ID/WG. 336/2, March 1981.

______ , *Second World-wide Study on the Petrochemical Industry: Process of Restructuring*, UNIDO, Turkey, ID/WG. 336/3, May 1981.

——— , *Second Consultation on the Petrochemical Industry*, UNIDO, Turkey, Report, June 1981.

——— , *Fibre Reinforced Composites*, UNIDO, Brazil, ID/WG. 368/8, April 1982.

——— , *Packaging and Plastics*, UNIDO, Brazil, ID/WG. 392/1, March 1983.

——— , *Plastics in the Building and Construction Industry*, UNIDO, Brazil, ID/WG. 392/4, March 1983.

——— , *World Demand for Petrochemical Products and the Emergence of New Producers from the Hydrocarbon Rich Developing Countries*. UNIDO, IS. 427, 19 December 1983.

——— , and Arni, V.R.S. (consultant), *Emerging Petrochemicals Technology: Implications for the Developing Countries*, UNIDO, IS. 350, October 1982.

——— , and Drewry, H.P. (shipping consultants), *Transport Costs for Shipping Petrochemicals, 1975–1985*, UNIDO, London, PC. 49, July 1982.

——— , and GOIC, *Petrochemical and other Industrial Uses of Natural Gas*, UNIDO, Bahrain, PC. 24, November 1981.

United Nations, *Report of the First United Nations Interregional Conference on the Development of Petrochemical Industries in Developing Countries*, Studies in Petrochemicals, U.N., Tehran, November 1966.

——— , *Yearbook of Industrial Statistics* (up to 1980), UN, New York (various years).

——— , 'The chemical industry', in *World Industry in 1980 – Restructuring World Industry: Trends and Prospects in Selected Industrial Branches*, UN, New York, 1981, pp. 111–32.

Vowles, C., 'The efficient use of energy through plastics', *Plastics Today*, No. 6, 1979, pp. 6–8.

Waddams, A.L., *Chemicals from Petroleum: An Introductory Survey*, 4th edn., Gulf, London, 1978.

Wallace, D.M., 'Saudi Arabia building costs', *Hydrocarbon Processing*, November 1976, pp. 189–96.

Walters, Sir Peter, 'Oil and petrochemicals – alchemy for all seasons?', *Chemistry and Industry*, 6 May 1985.

Wei, J. *et al.*, *The Structure of the Chemical Processing Industries: Function and Economics*, McGraw Hill, New York, 1979.

Wett, T., 'How petrochemicals changed the world', *The Oil and Gas Journal*, August 1977, pp. 425–8.

——— , 'The refiner's role in petrochemicals is growing', *The Oil and Gas Journal*, 3 April 1978, pp. 65–7.

Woodward, V.H., 'A disaggregated simulation model of the UK electrical engineering industry', in Gielnik, S.J. and Gossling, W.F. (eds), *Input–Output and Marketing*, Input–Output Publishing Co., London, 1980.

APPENDIX A:

STATISTICAL DATA FOR CHAPTER 1

Table A1 Downstream uses for selected petrochemical products (estimates for Western Europe)

		%		%
Ethylene	LDPE	40	HDPE	15
	EDC (for PVC)	19	Ethylene oxide	13
	Ethylbenzene	7	Other uses	6
Propylene	Polypropylene	26	Acrylonitrile	17
	Oxo alcohols	17	Propylene oxide	12
	Cumene	8	Other uses	20
Benzene	Ethylbenzene	49	Cumene	18
	Cyclohexane	11	Nitrobenzene	7
	Other uses	15		
Ethylene oxide	Ethylene glycol	45	Ethoxylates	21
	Glycol ethers	11	Ethanolamines	8
	Other uses	15		
Ethylene glycol	Polyester	50	Antifreeze	35
	Other uses	15		
LD Polyethylene	Film	74	Injection moulding	7
	Coatings	6	Cables	4
	Pipes	3.5	Blow moulding	3
Propylene oxide	Polyether polyols	65	Propylene glycol	25
	Other uses	10		
Polypropylene	Moulding	45	Fibre	37
	Film	10	Other uses	8
PVC	Pipes and fittings	28	Rigid profiles	12
	Wires and cables	10	Flexible films	10
	Rigid foil	8	Bottles	7
	Floor coverings	6	Coated fabrics	5
Styrene	Polystyrene	65	SB/SBR	13
	ABS	10	Polyester	8

Source: Chemical Age, 10 April 1981, p. 35.

Table A2 Production of ethylene (volume): trend from 1960 to 1975, forecasts for 1980 and 1985

(a) Percentage changes at annual rates

	1960/65	*1965/70*	*1970/73*	*1973/74*	*1974/75*	*1975/76*	*1976/80*	*1980/85*
EEC	(22.4)	24.7	17.1	7.9	−25.4	30.0		
Scandinavia	31.4	26.5	29.1	18.9	−12.1	19.4	(4)	(5)
Other European member countries			18.9	16.5	6.7	20.2		
United States	11.9	14.1	6.5	7.0	−14.2	7.4	6.5	6
Japan	57.7	31.8	10.4	0.1	−15.1	9.0	(5)	(5)
Sub-total (Western Europe, North America, Japan)	(17.2)	(19.6)	11.0	6.1	−18.6	16.0	(5)	(5.5)
Rest of world*	. .	. .	. .	. .	. .	. .	. .	(8.5)
Total*	. .	. .	. .	. .	. .	. .	. .	(6)

Table A2 *cont.*

(b) Total production (million tonnes)

	1960	*1965*	*1970*	*1973*	*1974*	*1975*	*1976*	*1980*	*1985*
EEC	(0.68)	1.87	5.63	9.04	9.75	7.27	9.45		
Scandinavia	0.01	0.05	0.16	0.35	0.42	0.37	0.44	(12.0)	(15.3)
Other European member countries	. .	. .	0.15	0.25	0.30	0.32	0.38		
United States	2.47	4.34	8.39	10.13	10.84	9.30	9.99	(13.5)	(18.0)
Japan	0.08	0.78	3.10	4.17	4.18	3.55	3.87	(4.7)	(6.1)
Sub-total (Western Europe, North America, Japan)	(3.3)	(7.3)	17.85	24.45	25.93	21.24	24.64	(30.2)	(39.4)
Rest of world*	—	—	—	—	—	(1.4)	—	(6.0)	(9.0)
Total*	—	—	—	—	—	(22.6)	—	(36.2)	(48.4)

* Excluding China and CMEA countries

( ) Estimates

. . Insignificant

Source: OECD, *The Petrochemical Industry*, Paris, 1979.

Table A3 Production of propylene (volume): trend from 1960 to 1975, forecasts for 1980 and 1985

(a) Percentage changes at annual rates

	1960/65	*1965/70*	*1970/73*	*1973/74*	*1974/75*	*1975/76*	*1976/80*	*1980/85*
EEC	(22.4)	23.1	16.2	5.2	−23.7	33.5		
Scandinavia	—	38.0	13.8	14.3	−28.1	39.1	(5)	(6)
Other European member countries	—	—	12.1	5.5	5.7	16.8		
United States	8.7	12.1	14.2	6.0	16.8	12.2	(8.75)	(8)
Japan	49*	27.8	9.5	−3.4	−10.6	14.3	(8.25)	(5)
Sub-total (Western Europe, North America, Japan)†	(16.5)	19.5	13.7	3.8	−18.2	20.3	(6)	(6.5)
Rest of world‡	. .	. .	. .	. .	. .	. .	. .	(3)
Total‡	. .	. .	. .	. .	. .	. .	. .	(6)

Table A3 *cont.*

(b) Total production (million tonnes)

	1960	*1965*	*1970*	*1973*	*1974*	*1975*	*1976*	*1980*	*1985*
EEC	(0.4)	1.09	3.08	4.83	5.08	3.87	5.17		
Scandinavia	—	0.02	0.03	0.06	0.10	0.05	0.11	(6.7)	(9.0)
Other European member countries	—	—	0.11	0.16	0.17	0.18	0.22		
United States	1.12	1.70	3.01	4.48	4.75	3.95	4.43	(6.2)	(9.1)
Japan	0.19*	0.63	2.15	2.82	2.73	2.44	2.69	(3.2)	(4.1)
Sub-total (Western Europe, North America, Japan)†	(1.6)	3.44	8.38	12.35	12.83	10.49	12.62	(16.1)	(22.2)
Rest of world‡	. .	. .	. .	. .	. .	(0.6)	. .	(2.8)	(3.2)
Total‡	. .	. .	. .	. .	. .	(11)	. .	(18.9)	(25.4)

* Million tonnes in 1962; change 1962–1965
† Excluding Canada
‡ Excluding China and CMEA countries
( ) Estimates
. . Insignificant

Source: As in Table A2.

Table A4 Production of butadiene (volume): trend from 1960 to 1975, forecasts for 1980 and 1985

(a) Percentage changes at annual rates

	1960/65	*1965/70*	*1970/73*	*1973/74*	*1974/75*	*1975/76*	*1976/80*	*1980/85*
EEC	(15)	(16.5)	15.1	6.9	−25.0	35.2		
Scandinavia	—	—	—	—	—	—	(4)	(4)
Other European member countries	—	—	—	55.6	—	7.1		
United States	7.5	2.9	5.4	1.1	−28.1	23.3	(5.75)	(3)
Japan	47.1	35.2	9.1	−1.9	−12.4	8.9	(5.25)	(5.25)
Sub-total (Western Europe, North America, Japan)	(9)	(10.5)	(9.75)	(2.75)	(−21.8)	24.2	(5)	(3.5)
Rest of world*	—	—	—	—	—	—	—	(4.5)
Total*	—	—	—	—	—	—	—	(3.5)

Table A4 *cont.*

(b) Total production (million tonnes)

	1960	*1965*	*1970*	*1973*	*1974*	*1975*	*1976*	*1980*	*1985*
EEC	(15)	(16.5)	15.1)	6.9	−25.0	35.2			
Scandinavia	—	—	—	—	—	—	(4)	(4)	
Other European member countries	—	—	—	55.6	—	7.1			
United States	7.5	2.9	5.4	1.1	−28.1	23.3	(5.75)	(3)	
Japan	47.1	35.2	9.1	−1.9	−12.4	8.9	(5.25)	(5.25)	
Sub-total (Western Europe, North America, Japan)	(9)	(10.5)	(9.75)	(2.75)	(−21.8)	24.2	(5)	(3.5)	
Rest of world*	. .	. .	. .	. .	. .	. .	. .	(4.5)	
Total*	—	—	—	—	—	. .	. .	(3.5)	. .

* Excluding China and CMEA countries

( ) Estimates

. . Insignificant

Source: As in Table A1.

Table A5 Production of benzene (volume): trend from 1960 to 1975, forecasts for 1980 and 1985

(a) Percentage changes at annual rates

	1960/65	*1965/70*	*1970/73*	*1973/74*	*1974/75*	*1975/76*	*1976/80*	*1980/85*
EEC	. .	. .	12.9	8.2	−29.8	36.5		
Scandinavia	—	—	—	—	—	—	(4)	(5)
Other European member countries	. .	. .	18.3	40.3	−23.8	42.9		
United States	12.6	6.4	8.6	4.1	−32.3	41.4	(7.5)	(5)
Japan	23.3	32.9	8.2	0.1	−19.5	16.8	(7)	(4.5)
Sub-total (Western Europe, North America, Japan)	—	—	(10.5)	(5.5)	(−29.5)	35.0	(6)	(5)
Rest of world*	. .	. .	. .	. .	. .	. .	. .	(9)
Total*	. .	. .	. .	. .	. .	. .	. .	(5.25)

Table A5 *cont.*

(b) Total production (million tonnes)

	1960	*1965*	*1970*	*1973*	*1974*	*1975*	*1976*	*1980*	*1985*
EEC	. .	. .	12.9	8.2	−29.8	36.5			
Scandinavia	—	—	—	—	—	—	(4)	(5)	
Other European member countries	. .	. .	18.3	40.3	−23.8	42.9			
United States	12.6	6.4	8.6	4.1	−32.3	41.4	(7.5)	(5)	
Japan	23.3	32.9	8.2	0.1	−19.5	16.8	(7)	(4.5)	
Sub-total (Western Europe, North America, Japan)	. .	. .	(10.5)	(5.5)	(−29.5)	35.0	(6)	(5)	
Rest of world*	. .	. .	. .	. .	. .	. .	. .	(9)	
Total*	. .	. .	. .	. .	. .	. .	. .	(5.25)	

* Excluding China and CMEA countries

( ) Estimates

. . Insignificant

Source: As in Table A2.

Table A6 Production capacity of ethylene: trend from 1970 to 1975, forecasts for 1980 and 1985

(a) Percentage rate of utilisation

	1970	*1971*	*1972*	*1973*	*1974*	*1975*	*1976*	*1980*	*1985*
EEC	88.6	87.7	87.7	89.5	86.2	61.8	78.6		
Scandinavia	78.6	80.8	35.5	98.6	95.9	78.1	91.3	(70)	(85)
Other European member countries	83.9	100	69.0	63.5	74.0	72.6	87.4		
United States	84.7	. .	. .	. .	96.3	73.6	. .	(78)	(85)
Japan	91.0	88.3	80.4	84.6	82.0	66.5	73.2	(80)	(85)
Sub-total (Western Europe, North America, Japan)	(87)	. .	. .	. .	. .	(68)	. .	(75)	(85)
Rest of world*	. .	. .	. .	. .	. .	(68)	. .	((75)	(85)
Total	. .	. . .	. .	. .	. .	(68)	. .	(75)	(85)

Table A6 *cont.*

(b) Total production (million tonnes)

	1970	*1971*	*1972*	*1973*	*1974*	*1975*	*1976*	*1980*	*1985*
EEC	6.36	7.03	8.57	10.11	11.32	11.77	12.03		
Scandinavia	0.21	0.26	0.34	0.36	0.44	0.48	0.49	(17.4)	(18.0)
Other European member countries	0.18	0.18	0.29	0.40	0.40	0.44	0.44		
United States	9.90	. .	. .	. .	11.26	12.64	. .	17.2	21.1
Japan	3.40	4.00	4.79	4.93	5.09	5.15	5.15	(5.8)	(7.0)
Sub-total (Western Europe, North America, Japan)	(20.5)	. .	. .	. .	. .	(31.0)	. .	(40.4)	(46.1)
Rest of world*	. .	. .	. .	. .	. .	(2)	. .	(8)	(10.5)
Total	. .	. .	. .	. .	. .	(33)	. .	(48.4)	(56.6)

* Excluding China and CMEA countries

( ) Estimates

. . Insignificant

Source: As in Table A2.

Table A7 Production capacity of propylene: trend from 1970 to 1975, forecasts for 1980 and 1985

(a) Percentage rate of utilisation

	1970	*1971*	*1972*	*1973*	*1974*	*1975*	*1976*	*1980*	*1985*
EEC	81.2	80.2	80.3	84.1	81.7	59.7	76.5		
Scandinavia	28.6	29.2	36.4	42.1	61.3	33.7	62.9	(69)	(85)
Other European member countries	88.0	100	78.2	72.1	80.9	78.3	91.5		
United States	. .	. .	. .	. .	. .	(65)	. .	(80)	(85)
Japan	—	—	—	—	—	(72)	—	(80)	(85)
Sub-total (Western Europe, North America, Japan)	. .	. .	. .	. .	. .	(64)	. .	(75)	(85)
Rest of world†	. .	. .	. .	. .	. .	(64)	. .	(75)	(85)
Total	. .	. .	. .	. .	. .	(64)	. .	(75)	(85)

Table A7 *cont.*

(b) Total production (million tonnes)

	1970	*1971*	*1972*	*1973*	*1974*	*1975*	*1976*	*1980*	*1985*
EEC	3.79	4.17	4.88	5.74	6.22	6.48	6.76		
Scandinavia	0.11	0.13	0.14	0.15	0.16	0.16	0.17	(9.7)	(10.6)
Other European member countries	0.13	0.13	0.17	0.22	0.22	0.24	0.24		
United States	. .	. .	. .	. .	. .	(6.1)	—	(7.8)	(10.7)
Japan	. .	. .	. .	. .	. .	3.5	3.5	(4.0)	(4.7)
Sub-total (Western Europe, North America, Japan)	. .	. .	. .	. .	. .	(16.5)	. .	(21.5)	(26.0)
Rest of world†	. .	. .	. .	. .	. .	(0.9)	. .	(3.7)	(3.8)
Total	. .	. .	. .	. .	. .	17.4	. .	(25.2)	(29.8)

* Excluding Canada
† Excluding China and CMEA countries
( ) Estimates
. . Insignificant

Source: As in Table A2.

Table A8 Production capacity of butadiene: trend from 1970 to 1975, forecasts for 1980 and 1985

(a) Percentage rate of utilisation

	1970	*1971*	*1972*	*1973*	*1974*	*1975*	*1976*	*1980*	*1985*
EEC	93.9	89.1	82.8	82.0	79.1	54.9	73.0		
Scandinavia	—	—	—	—	—	—	—	(72)	(83)
Other European member countries	62.5	87.5	80.0	69.2	87.5	87.5	93.8		
United States	. .	. .	. .	. .	. .	(65)	. .	(80)	(85)
Japan	100	76	93	93	85	70	76	(80)	(85)
Sub-total (Western Europe, North America, Japan)	. .	. .	. .	. .	. .	(62)	. .	(78)	(86)
Rest of world	. .	. .	. .	. .	. .	(66)	. .	(80)	(84)
Total*	. .	. .	. .	. .	. .	(63)	. .	(78)	(86)

Table A8 *cont.*

(b) Total production (million tonnes)

	1970	*1971*	*1972*	*1973*	*1974*	*1975*	*1976*	*1980*	*1985*
EEC	0.91	0.99	1.24	1.60	1.77	1.92	1.95		
Scandinavia	—	—	—	—	—	—	—	(2.36)	(2.4)
Other European member countries	0.01	0.01	0.01	0.03	0.03	0.03	0.03		
United States	. .	. .	. .	. .	. .	(1.85)	. .	(2.55)	(2.8)
Japan	0.50	0.76	0.70	0.70	0.75	0.82	0.82	(0.9)	(1.1)
Sub-total (Western Europe, North America, Japan)	. .	. .	. .	. .	. .	(4.8)	. .	(5.8)	(6.3)
Rest of world	. .	. .	. .	. .	. .	(0.3)	. .	(0.5)	(0.6)
Total*	. .	. .	. .	. .	. .	(5.1)	. .	(6.3)	(6.9)

* Excluding China and CMEA countries

( ) Estimates

. . Insignificant

Source: As in Table A2.

Table A9 Production capacity and utilisation rates of benzene: trend from 1970 to 1975, forecasts for 1980 and 1985

(a) Percentage rate of utilisation

	1970	*1971*	*1972*	*1973*	*1974*	*1975*	*1976*	*1980*	*1985*
EEC	73.5	77.4	72.8	82.7	77.7	52.4	69.4		
Scandinavia									
Other European member countries	40.8	45.1	63.2	64.9	84.2	64.2	81.5		
United States	85	. .	. .	. .	86	56	(75)	(80)	(85)
Japan	85	. .	. .	. .	. .	67	(65)	(80)	(85)
Sub-total (Western Europe, North America, Japan)	(7.9)	. .	. .	. .	(83)	(57)	(71)	(76)	(85)
Rest of world*	. .	. .	. .	. .	. .	(60)	. .	(76)	(85)
Total	. .	. .	. .	. .	. .	(60)	. .	(76)	(85)

Table A9 *cont.*

(b) Total production (million tonnes)

	1970	*1971*	*1972*	*1973*	*1974*	*1975*	*1976*	*1980*	*1985*
EEC	3.68	3.75	4.33	4.70	5.41	5.64	5.81		
Scandinavia	—	—	—	—	—	—	—	(7.3)	(7.4)
Other European member countries	0.21	0.21	0.22	0.22	0.24	0.24	0.27		
United States	4.4	. .	. .	. .	5.8	6.1	(6.4)	(8.4)	(10.1)
Japan	1.85	. .	. .	. .	2.4	2.4	(2.6)	(3.1)	(3.6)
Sub-total (Western Europe, North America, Japan)	(10.4)	. .	. .	. .	(14.1)	(14.7)	(15.7)	(18.8)	(21.1)
Rest of world*	. .	. .	. .	. .	. .	(0.5)	. .	(2.0)	(2.7)
Total*	. .	. .	. .	. .	. .	(15.2)	. .	(20.8)	(23.8)

* Excluding China and CMEA countries
( ) Estimates
. . Insignificant

Source: As in Table A2.

Table A10 Representative examples of weight reduction in conversion from metal to plastic

Application	*Plastic*	*Material replaced*	*% weight reduction*
Radiator fan	Polypropylene	Sheet metal assembly	30
Exhaust gas cycle valve	PPS	Welded sheet metal	80
Brake piston	Phenolic	Steel	56
Door window bracket	Acetal	Welded steel stamping	72
Steering column housing	Glass/nylon	Die cast zinc	66
Oil pan	Glass/nylon	Steel	50
Seat shell	Glass/polypropylene	Steel	50
Seat back/load floor	Polypropylene	Steel	50
Front end retained (used with RIM fascia)	Glass/polypropylene	Steel	66
Gasoline tank	Polyethylene	Steel	40
Gasoline tank	Polyethylene	Steel	31
Mirror housing	Mineral/nylon	Die cast zinc	5
Bumper bar	Glass/polypropylene	Steel	30
Leaf spring	SMC	Steel	82
Drive shaft	Glass/graphite/epoxy	Steel	57
Brake pedal arm	Glass/epoxy	Steel	53
Bumper energy attenuator	SMC	Steel/hydraulic mechanism	57
Bumper bar	SMC	Steel	54
Fender extender	Mineral/nylon	Die cast zinc	1.5
Air compressor bracket	Glass/graphite/ polyester	Cast iron	70
Fender	RIM urethane	Steel	60
Wheels	SMC	Steel	40
Bellows	Polyurethane	Rubber	50
Door	SMC	Steel	43
Headlamp	Polycarbonate	Glass	65

Source: Jim Best, Market Research Inc., Society of Plastics Engineers 39th Annual Technical Conference and Exhibition, 1981.

Table A11 Energy saved over life cycle of the car by replacing metal with one pound of plastic

	Pounds of gasoline equivalent saved			
One-pound plastic part	*Replace die cast aluminium*	*Replace sheet aluminium*	*Replace steel*	*Replace die cast zinc*
Low density polyethylene	2.2	3.4	5.5	6.7
PVC	2.0	3.2	5.3	6.5
Glass/polystyrene	1.9	3.1	5.2	6.4
Glass/RIM urethane	1.9	3.1	5.2	6.4
RIM urethane	1.8	3.0	5.1	6.3
High-density polyethylene	1.8	3.0	5.1	6.3
Polystyrene	1.7	3.0	5.1	6.3
SMC	1.7	3.0	5.1	6.3
Glass/ABS	1.7	3.0	5.1	6.3
Filled polypropylene	1.7	2.9	5.0	6.2
ABS	1.5	2.7	4.6	6.0
Glass/PBT polyester	1.2	2.4	4.5	5.7
Filled nylon 66	1.2	2.4	4.5	5.7
Glass/polypropylene	1.1	2.4	4.5	5.7
Glass/polycarbonate	1.1	2.3	4.4	5.6
PBT polyester	0.7	2.0	4.1	5.3
Polypropylene	0.7	1.9	4.0	5.2
Polycarbonate	0.6	1.8	3.9	5.1
Glass/nylon 66	0.5	1.8	3.9	5.1
Acrylic	0.4	1.7	3.8	5.0
Nylon 66	(0.2)	1.0	3.1	4.3
Nylon 6	(0.4)	0.8	2.9	4.1
Acetal	(0.8)	0.4	2.5	3.7

Note: 1 pound gasoline = 0.615 litres

Source: As in Table A10.

APPENDIX B:

STATISTICAL DATA FOR CHAPTER 2

Table B1 The effect of plant capacity on capital investment and production costs in the chemical industries of the developed countries in the 1960s

Production designation	*Capacity, thousands of tons/year*	*Unit capital investment*		*Production costs*	
		US$/ton	*%*	*US$/ton*	*%*
Ammonia* (from natural gas)	36	139	100.0	46.0	100.0
	102	108	77.7	38.0	82.6
	180	89	64.0	34.0	74.0
Butadiene*	10	600	100.0	239.0	100.0
	20	450	75.0	202.0	84.5
	40	338	56.0	178.0	74.4
Ethylene† (by-products based on the price of the chemical raw material)	50	150	100.0	94.8	100.0
	100	120	80.3	70.3	74.1
	150	100	66.6	66.1	69.7
	300	90	60.0	47.2	49.8
	454	77	51.0	42.8	45.1
Polyvinyl chloride*	6	285	100.0	290.0	100.0
	20	170	60.0	250.0	86.2
	40	129	46.0	239.0	82.4
Styrene‡	12	275	100.0	180.3	100.0
	48	162	58.9	149.6	83.0
	96	116	42.1	140.0	77.8
Polystyrene‡	10	278	100.0	235.0	100.0
	40	181	65.3	210.0	89.0
	80	156	56.3	202.0	86.0

* Calculated on the basis of documents produced at the Seventh Petroleum Congress of Arab Countries held in March 1970 in Kuwait.

† Study of Feedstock and Process in the Petrochemical Industry, UNIDO, 1969, page 252.

‡ Studies in the Development of the Plastics Industries (United Nations Publication, Sales No. E.69.II.B.25), pp. 43–9.

Source: UNIDO, 1978, p. 79.

Table B2 Trends of export prices for specific chemical products, oil, naphtha and gas (US$ per ton)

Product	*1970*	*1972*	*1974*	*1975*	*1976*	*1977*	*1978*	*1979*	*1980*
Ethylene	70–90	80–90	260–285	260–330	240–330	295–315	286–370	310–590	410–740
Ammonia	35–50	38–45	135–150	150–230	105–123	100–120	95–110	120–160	140–200
Methanol	60–90	50–70	100–250	100–150	100–130	90–135	120–130	150–175	200–240
High-density polyethylene	290–370	270–340	700–790	615–660	620–680	630–660	580–700	770–1000	950–1200
Low-density polyethylene	230–300	250–300	680–740	550–600	550–600	500–560	515–560	840–950	980–1150
Polyvinyl chloride	290–330	220–380	650–760	510–570	520–610	510–580	580–700	840–950	870–1100
Oil	13.31	18.19	85.41	85.01	91.25	100.72	100.72	102.80	
Naphtha	16.08	20.02	123.25	109.73	130.69	125.12	146.14		
Natural gas (for 000 m^3)	9.37	10.81	18.31	39.20	60.39	64.80	76.79	87.16	156

Source: UNIDO, ID/WG. 336/3. May 1981, p. 84.

Table B3 Typical production costs of petrochemicals

	Capacity 10^3 ton/year	Fixed capital cost US$ m.	Manufacturing cost 10^3 US$/year							Product cost ($/ton)
			Raw materials	Utilities	Catalysts chemicals	Manpower	Other charges	Amortisation and return	Total manufacturing cost	
Methanol	200	44	21,120	300	300	750	2,860	8,360	33,690	168
Vinyl chloride	150	50	37,560	4,515	1,900	1,065	3,250	9,500	57,790	385
Styrene	150	55	44,930	6,390	1,460	975	3,575	10,450	67,780	452
Caprolactam	80	120	18,400	9,624	12,000	1,215	7,800	22,800	71,839	764*
DMI	60	60	16,170	3,610	330	1,065	3,900	11,400	36,475	608
TPA	80	78	20,636	5,700	400	1,065	5,070	14,820	47,691	596
Ethylene oxide	80	27	25,600	1,328	200	750	1,755	5,130	34,763	435
Acrylonitrile	100	71	25,960	6,300	3,500	1,110	4,615	13,490	54,975	550
PVC	70	30	28,030	820	300	2,070	1,950	5,700	38,870	555
Hd polyethylene	70	65	23,700	2,620	1,100	1,395	4,225	12,350	45,290	647
Ld polyethylene	110	86	36,600	3,190	1,920	1,395	5,590	16,340	65,035	591
Polystyrene	50	30	22,600	620	200	1,920	1,950	5,700	32,990	660
Polyester fibres	12	20	8,894	50	400	3,750	1,300	3,180	17,574	1,460
Nylon fibres	10	25	8,404	59	300	4,500	1,620	4,750	19,633	1,964
Acrylic fibres	10	16	5,610	59	400	3,000	1,040	3,040	13,149	1,315
Polybutadiene	40	25	14,800	1,368	1,500	1,575	1,625	4,750	26,618	640
SBR	60	18	26,380	2,232	1,700	1,875	1,170	3,420	36,777	613

* Taking into consideration a by-products valorisation amounting to 10.7 million/year.

Source: UNIDO, 1978, p. 50.

Table B4 Total unit investment costs for selected petrochemicals and intermediates in the United States

	Small plants		*Large plants*	
Intermediate/product	*Individual plant ($/ton/year)*	*Total investment* ($/ton/year)*	*Individual plant ($/ton/year)*	*Total investment* ($/ton/year)*
Ethylene	802†	—	611*	—
HDPE	636	1,450	448	1,094
LDPE	1,000	1,850	692	1,340
LLDPE	634	1,377	461	1,027
Ethylene oxide	1,005	1,773	701	1,286
Ethylene glycol	234	1,556	153	1,112
Ethyl benzene	112	328†	77	242†
Styrene	282	658	215	493
Polystyrene	487	1,158	352	855
SBR	1,331	1,478	856	966
DMT:	1,181†	—	883†	—
PET	1,178	2,919	828	2,150
TPA:	1,117†	—	865†	—
PET	1,116	2,632	694	1,835
Chlorine	661	—	451	—
VCM	414	1,195	312	875
PVC	998	2,000	645	1,514

* Including upstream investment
† Excludes investment for feedstock extraction plant

Source: UNIDO, ID/WG. 336/3, May 1981, p. 105.

Table B5 Impact of plant size on transfer prices (United States Gulf Coast)

Product	*Small plant (100)*[‡] *$/ton*	*Medium plant (200)*[‡] *$/ton*	*Large plant (400)*[‡] *$/ton*
DMT	1,392	1,265	1,195
Ethyl benzene	799	788	775
Ethylene-propylene*	685	613	582
Ethylene-propylene butadiene-benzene[†]	862	772	732
Ethylene glycol	768	740	720
Ethylene oxide	1,075	965	902
HDPE	1,135	1,061	1,025
LDPE	1,079	979	913
LLDPE	1,006	951	915
Methanol	314	288	274
Polyethylene terephthalate (PET)	1,924	1,741	1,620
Polypropylene	1,067	986	938
Polystyrene	1,146	1,068	1,043
PVC	1,130	1,090	1,043
Styrene	920	893	885
SBR	2,368	2,079	1,913
Terephthalic acid (TPA)	1,294	1,207	1,145
VCM	836	798	769

* Ethylene price using ethane-propane feedstock
† Ethylene price using naphtha feedstock
‡ Figures in parentheses are in units of thousand tonnes/yr i.e. 100 x 10^3 tonne/yr

Source: As in Table B4.

Table B6 Refinery yields on crude oil (1982)

Products	*N. America*	*W. Europe*	*Japan*
		(% by weight)	
Gasoline	44	24	21
Middle distillates	29	36	33
Fuel oil	10	24	37
Other	17	16	9
Total	100	100	100

Source: BP *Statistical Review of World Energy*, 1982, p. 16.

Table B7 Capital and exploration expenditures by oil companies ($ m.)

	1970	*1971*	*1972*	*1973*	*1974*	*1975*	*1976*	*1977*	*1978*	*1979*	*1980*
World											
Crude oil and natural gas	6,650	6,520	9,590	12,415	18,765	18,295	23,860	28,680	33,675	44,500	61,300
Natural gas liquids plants	580	695	515	510	770	960	1,915	3,780	4,030	4,565	5,575
Pipelines	850	1,200	1,230	1,230	2,460	5,995	7,575	6,660	5,780	5,775	6,475
Tankers	2,575	2,875	3,775	6,550	8,900	9,240	8,675	3,700	2,950	2,250	3,400
Refineries	4,000	4,755	4,955	4,865	7,720	8,725	6,910	8,290	10,675	11,775	13,475
Chemical plants	1,525	1,535	1,350	1,175	1,995	3,145	4,500	6,375	6,650	7,235	8,300
Marketing	3,220	3,380	2,825	2,480	2,215	2,160	2,180	2,670	3,240	3,750	5,050
Other	725	840	710	770	875	1,105	1,110	1,425	1,375	1,775	1,950
Total capital expenditures	20,125	21,800	24,950	29,995	43,700	49,625	56,725	61,580	68,375	81,625	105,525
United States											
Crude oil and natural gas	4,110	3,185	5,740	7,290	11,225	9,055	13,135	14,855	17,325	24,150	32,350
Natural gas liquids plants	225	200	175	150	225	325	325	325	325	450	750
Pipelines	450	550	300	450	1,400	3,500	3,625	1,800	750	1,100	1,550
Tankers	100	125	125	100	200	225	275	500	600	350	600
Refineries	1,075	1,050	900	1,050	1,775	2,100	1,575	1,200	1,750	2,625	3,250
Chemical plants	550	500	450	425	825	1,500	2,200	2,500	2,300	2,000	2,400
Marketing	1,450	1,350	1,100	850	650	650	625	775	950	975	1,300
Other	265	290	260	325	325	370	325	445	475	550	700
Total capital expenditures	8,225	7,250	9,050	10,640	16,625	17,725	22,085	22,400	24,475	32,200	42,100

Table B7—*cont.*

	1970	*1971*	*1972*	*1973*	*1974*	*1975*	*1976*	*1977*	*1978*	*1979*	*1980*
Western Europe											
Crude oil and natural gas	300	500	650	1,300	2,375	3,600	4,200	5,275	5,975	7,200	10,500
Natural gas liquids plants	50	75	50	25	40	50	50	225	200	350	400
Pipelines	75	150	400	350	475	1,000	1,100	875	950	700	550
Refineries	1,050	1,400	1,500	1,550	2,250	2,350	1,450	2,000	2,500	2,525	2,950
Chemical plants	725	800	800	600	850	1,100	1,200	1,300	1,500	1,600	1,800
Marketing	900	1,000	825	800	750	700	725	875	1,100	1,400	2,125
Other	160	225	225	200	180	250	225	325	250	300	300
Total capital expenditures	3,260	4,150	4,450	4,825	6,920	9,050	8,950	10,875	12,475	14,075	18,625
Middle East											
Crude oil and natural gas	275	450	500	850	975	1,000	1,375	1,925	1,550	1,700	2,450
Natural gas liquids plants	5	25	10	5	25	25	675	1,400	1,500	1,650	2,500
Pipelines	75	65	140	130	120	350	1,250	2,150	1,400	1,550	2,200
Refineries	140	125	225	300	450	400	925	1,250	2,050	2,100	2,350
Chemical plants	25	40	25	30	40	50	150	900	1,000	1,200	1,750
Marketing	25	25	25	25	25	25	25	50	50	100	125
Other	20	95	50	50	135	175	275	125	100	125	150
Total capital expenditures	565	825	975	1,390	1,770	2,025	4,675	7,800	7,650	8,425	11,525

Source: Chase Manhattan Bank, 1980.

Table B8 Petrochemicals control in the EEC region (per cent)

	Year	*Ethylene*	*Butadiene*	*Benzene*	*LDPE*	*HDPE*	*PVC*	*PP*	*OX. Ethyl*	*Styrene*
MpEEC*	1976	25.9	39.6	22.7	30.1	18.4	13.6	22.3	20.8	25.8
	1980	23.8	37.6	24.4	28.3	20.4	15.7	23.2	31.0	24.7
	1983	21.5	38.1	24.4	28.0	22.2	18.7	22.8	30.4	26.7
McEEC	1976	26.3	25.4	16.9	33.4	59.0	57.8	71.5	55.6	36.4
	1980	26.5	26.1	16.1	30.3	53.5	51.4	54.8	53.5	27.2
	1983	26.9	24.0	16.2	30.0	47.8	49.6	51.2	17.0	26.5
MpUS†	1976	11.4	9.2	20.3	—	3.5	1.7	—	—	8.0
	1980	12.1	8.5	17.2	4.6	3.0	1.6	5.5	—	5.8
	1983	11.2	8.5	17.2	4.6	2.9	2.6	5.0	—	5.7
McUS	1976	7.2	8.5	9.4	8.8	3.5	5.0	—	11.6	14.7
	1980	6.8	7.5	11.2	7.4	3.2	4.8	5.0	—	21.7
	1983	6.3	7.5	11.2	7.3	3.1	3.0	4.5	—	21.1
Ipc†	1976	29.2	17.3	30.7	27.7	15.6	22.0	6.2	11.9	15.1
	1980	30.7	20.8	30.8	29.4	19.8	23.5	11.5	15.5	20.5
	1983	31.1	21.6	30.8	30.1	24.0	24.1	16.5	22.6	20.0

* These are the percentages of the total EEC capacities controlled on a pro rata basis fo their financial participations by each group.

† MpEEC, Oil majors of the EEC (BP, Shell, CFP); MpUS, Oil majors in the United States (Gulf, Exxon, etc.); Ipc: independent petrochemical producers.

Sources: UNIDO, ID/WG. 336/3, May 1981; *Oil and Gas Journal*, September issues, various years.

Table B9 Control of the petrochemical industry in the United States (percentages of production capacities)

		Ethylene	*Benzene*	*HDPE*	*LDPE*	*PP*
Oil companies						
US	1976	32.6	na	21.9	22.5	38.1
	1980	42.1	38.1	37.1	28	42.6
	1983	40.1	37.9	32.6	27.6	41.7
EEC	1976	9.9	na	—	—	9.3
	1980	10.6	8.5	—	—	10.9
	1983	13.7	7.5	—	—	10.5
Chemical companies						
US	1976	38.2	na	32.3	44.4	38.4
	1980	30.7	9	26.7	37.2	25.6
	1983	30.4	12.5	22.4	40.1	28.2
EEC	1976	1.7	na	12.1	—	—
	1980	2.7	—	10.9	—	3.6
	1983	2.5	1.5	15.1	—	3.5
Others						
	1976	18.2	na	33.7	33.1	14.2
	1980	13.9	44.4	26.3	34.7	17.3
	1983	13.3	40.6	29.9	32.2	16.1

Source: As in Table B8.

Table B10 Western European links between olefins producers, oil refiners and aromatics producers (1980)

Company/grouping	*Refining*	*Olefins*	*Benzene*	*PX*	*OX*
ANIC/ENI	X	X	X	X	X
BASF/ROW/Wintershall	X	X	X	—	—
BP/Erdölchemie/ Naphthachemie	X	X	X	X	X
British Celanese	—	X	—	—	—
CdF Chimie	X	X	X	—	—
Calter/Texaco	X	X	X	—	X
CNP/Petrogal	X	X	X	X	X
Dow Chemical	X	X	X	—	—
DSM	—	X	X	—	—
Elf/ATO/CFP	X	X	X	X	X
Enpetrol	X	X	X	—	—
Exxon	X	X	X	—	—
Gulf	X	X	XX	—	—
ICI/Phillips	X	X	X	X	—
IQA	X	X	—	—	—
Marathon	X	X	X	—	—
Montedison	X	X	X	X	X
Neste Oy	X	X	X	—	—
Noretyl	X	X	—	—	—
MOV	X	X	—	—	—
Petrochim (Phillips/Fina)	X	X	X	—	—
SIR/Runianca	X	X	X	—	—
Shell	X	X	X	X	X
Solvay	X	X	—	—	—
Veba-Huels	X	X	X	X	X
UREK	X	X	X	X	X

Source: Chem-Systems, London, 1981.

Table B11 Japanese links between oil refiners and aromatics producers (1980)

Company/grouping	*Refining*	*Olefins*	*Benzene*	*PX*	*OX*
Asia Oil	X	—	X	—	—
Idemitsu	X	X	X	—	—
Kawatetsu	—	—	X	—	—
Kuraray	—	—	—	X	—
Maruzen	X	—	X	—	X
Matsuyama	—	—	—	X	—
Mitsubishi Group	X	X	X	X	X
Mitsui Group	—	X	X	—	—
Nippon Petrochemical	—	X	X	—	—
Osaka Petrochemical	X	X	X	—	—
Sanyo Petrochemical	X	X	X	—	—
Shin Daikyowa	X	X	X	—	—
Sumitomo Group	X	X	X	—	—
Toa Oil	X	—	X	—	—
Tonen Petrochemical	X	X	X	X	X
Toray	—	—	X	X	X

Source: Chem-Systems, London, 1981.

APPENDIX C:

STATISTICAL DATA FOR CHAPTER 3

Table C1 The dependence of the gross production of synthetic resins and plastics materials industry (29) on final demand components and the supporting industries (United Kingdom, 1974).

		Final demand (£ million)					
SIC No.	*Industry*	*105 Consumers' expenditure*	*106 General government consumption*	*107 Fixed capital formulation*	*108 Stock building*	*109 Export goods and services*	*110 Total final output*
1	Agriculture	5.12	0.16	0.14	0.26	0.36	6.04
2	Forestry and fishing	0.30	0.01	0.00	0.06	1.10	0.49
3	Stone, slate, etc. extraction	0.02	0.00	0.00	0.04	0.17	0.23
4	Other mining and quarrying	0.00	0.05	0.00	0.00	0.09	0.14
5	Water supply	0.73	0.05	0.17	0.00	0.00	0.95
6	Gas	1.51	0.10	0.11	0.00	0.00	1.73
7	Electricity	2.70	0.38	0.73	0.00	0.00	3.81
8	Coal-mining	0.50	0.04	0.02	0.08	0.05	0.54
9	Petroleum and natural gas extraction	0.00	0.00	0.01	0.00	0.00	0.01
10	Coke and manufactured fuel	0.17	0.03	0.00	0.06	0.09	0.24
11	Mineral oil refining, etc.	0.30	0.11	0.00	0.11	0.54	1.07
12	Grain milling	0.50	0.01	0.00	0.01	0.03	0.55
13	Bread, etc.	11.82	0.23	0.01	0.05	0.40	12.40
14	Meat and fish products	7.00	0.25	0.00	0.03	0.79	8.07
15	Milk and milk products	5.93	0.28	0.00	0.18	0.27	6.66

Table C1—*cont.*

		Final demand (£ million)					
SIC No.	*Industry*	*105 Consumers' expenditure*	*106 General government consumption*	*107 Fixed capital formulation*	*108 Stock building*	*109 Export goods and services*	*110 Total final output*
16	Sugar	0.24	0.01	0.00	0.00	0.13	0.38
17	Cocoa, chocolate, etc.	4.89	0.13	0.02	0.09	0.93	6.06
18	Animal foods	0.63	0.00	0.00	0.04	0.06	0.74
19	Oils and fats	0.11	0.00	0.00	0.02	0.04	0.18
20	Other food	9.65	0.20	0.01	0.12	0.69	10.44
21	Soft drinks	1.79	0.05	0.00	0.09	0.08	2.00
22	Alcoholic drinks	3.24	0.01	0.03	0.18	1.42	4.66
23	Tobacco	5.06	0.01	0.01	0.21	0.77	5.64
24	General chemicals	0.33	0.24	0.16	0.49	6.57	7.60
25	Pharmaceutical chemicals	1.04	1.85	0.04	0.23	2.29	5.46
26	Toilet preparations	6.32	0.04	0.00	0.20	1.49	8.04
27	Paint	7.41	0.19	0.03	0.65	3.95	12.23
28	Soap and detergents	1.85	0.11	0.01	0.00	0.83	2.80
29	Synthetic resins, etc.	3.21	1.61	1.51	24.80	327.24	362.48
30	Dyestuffs and pigments	0.01	0.01	0.03	0.12	1.05	1.22

Table C1—*cont.*

		Final demand (£ million)					
SIC No.	*Industry*	*105 Consumers' expenditure*	*106 General government consumption*	*107 Fixed capital formulation*	*108 Stock building*	*109 Export goods and services*	*110 Total final output*
31	Fertilizers	0.13	0.05	0.02	0.01	0.71	0.90
32	Other chemical industries	4.36	1.49	0.17	0.13	9.71	15.86
33	Iron castings, etc.	0.01	0.00	0.08	0.11	0.16	0.36
34	Other iron and steel	0.00	0.01	0.12	0.10	1.59	1.81
35	Aluminium and alloys	0.01	0.00	0.01	0.03	0.29	0.34
36	Other non-ferrous metals	0.01	0.00	0.01	0.06	0.92	1.00
37	Agricultural machinery	0.05	0.00	0.45	0.02	0.32	0.84
38	Machine tools	0.00	0.02	0.84	0.03	0.72	1.61
39	Pumps, etc.	0.01	0.03	0.51	0.05	1.07	1.67
40	Industrial engines	0.00	0.04	0.07	0.01	0.31	0.44
41	Textile machinery	0.02	0.00	0.25	0.02	0.86	1.15
42	Construction equipment	0.00	0.03	1.23	0.04	1.73	3.03
43	Office machinery	0.08	0.06	0.31	0.09	0.80	1.34
44	Other non-electrical machinery	0.12	0.08	1.85	0.27	3.25	5.53
45	Industrial plant and steel work	0.01	0.17	3.64	0.17	1.60	5.59

Table C1—*cont.*

		Final demand (£ million)					
SIC No.	*Industry*	*105 Consumers' expenditure*	*106 General government consumption*	*107 Fixed capital formulation*	*108 Stock building*	*109 Export goods and services*	*110 Total final output*
46	Other mechanical engineering	0.04	0.59	0.21	0.11	1.40	2.38
47	Instrument engineering	0.29	1.81	3.29	0.43	5.83	11.65
48	Electrical machinery	0.00	0.30	2.32	0.07	2.35	5.04
49	Insulated wires and cables	0.03	0.11	0.45	0.44	4.99	5.14
50	Telegraph and telephone equipment	0.00	0.11	4.78	0.09	0.99	5.99
51	Radio and electronic components	0.27	1.17	0.26	0.53	5.28	7.51
52	TV, radio, etc. equipment	5.09	0.58	4.24	0.07	1.56	11.55
53	Electronic computers	0.00	0.10	1.51	0.19	1.61	3.40
54	Radio and electronic capital goods	0.01	0.62	0.75	0.20	0.83	2.40
55	Domestic electrical appliances	8.23	0.28	0.91	0.26	3.50	13.17
56	Other electrical goods	0.48	0.47	0.12	0.07	2.54	3.68
57	Shipbuilding and marine engineering	0.01	1.13	1.11	0.04	0.87	3.08
58	Wheeled tractors	0.01	0.02	0.38	0.03	1.65	2.10
59	Motor vehicles	1.88	0.70	9.83	0.16	11.03	23.27
60	Aerospace equipment	0.00	1.81	0.03	0.36	2.05	4.26

Table C1—*cont.*

SIC No.	Industry	Final demand (£ million)					
		105 Consumers' expenditure	*106 General government consumption*	*107 Fixed capital formulation*	*108 Stock building*	*109 Export goods and services*	*110 Total final output*
61	Other vehicles	0.08	0.03	0.23	0.02	0.34	0.66
62	Engineers' small tools	0.01	0.01	0.11	0.00	0.18	0.31
63	Cutlery, etc.	0.24	0.00	0.02	0.01	0.83	1.09
64	Bolts, nuts, screws, etc.	0.02	0.01	0.02	0.00	0.16	0.21
65	Wire manufactures	0.03	0.02	0.01	0.02	0.00	0.16
66	Cans and metal boxes	0.01	0.01	0.01	0.05	0.14	0.21
67	Other metal goods	0.99	0.27	0.61	0.05	2.05	3.96
68	Production of synthetic fibres	0.46	0.04	0.28	0.43	12.29	13.50
69	Cotton, etc., weaving	0.91	0.04	0.00	0.16	1.57	2.69
70	Woollen and worsted	0.35	0.03	0.00	0.01	1.72	2.09
71	Hosiery and knitted goods	4.06	0.04	0.00	0.22	1.72	6.06
72	Carpets	9.65	0.30	0.05	0.59	3.21	12.62
73	Household textiles	1.06	0.10	0.00	0.04	0.24	1.36
74	Textile finishing	0.01	0.00	0.06	0.01	0.00	0.09
75	Other textiles	2.10	0.18	0.00	0.16	3.11	5.24

Table C1—*cont*.

		Final demand (£ million)					
SIC No.	*Industry*	*105 Consumers' expenditure*	*106 General government consumption*	*107 Fixed capital formulation*	*108 Stock building*	*109 Export goods and services*	*110 Total final output*
76	Leather goods and fur	1.18	0.01	0.00	0.00	1.80	2.99
77	Clothing	8.51	0.38	0.00	0.30	1.61	10.41
78	Footwear	9.02	0.19	0.01	0.01	2.14	11.37
79	Bricks, fireclay, etc.	0.05	0.00	0.01	0.03	0.16	0.26
80	Pottery and glass	0.60	0.08	0.11	0.08	1.62	2.49
81	Cement	0.03	0.00	0.00	0.01	0.05	0.10
82	Other building materials, etc.	0.20	0.53	0.00	0.00	0.25	0.99
83	Furniture and bedding, etc.	5.08	0.65	0.33	0.06	0.72	6.73
84	Timber and wool manufacture	0.67	0.17	1.69	0.58	0.23	3.35
85	Paper and board	0.09	0.35	0.00	0.19	1.51	2.15
86	Packaging products	0.23	0.22	0.01	0.20	0.75	1.41
87	Other paper and board products	1.75	0.96	0.00	0.21	0.91	3.41
88	Printing and publishing	2.90	0.47	0.01	0.05	1.45	4.88
89	Rubber	6.22	0.63	0.26	0.15	13.80	21.26
90	Plastics products	14.83	1.34	0.00	1.53	12.68	30.46

Table C1—*cont.*

91	Other manufacturing	8.91	2.06	0.27	0.50	10.20	21.93
92	Construction	11.42	7.85	75.99	1.06	0.89	97.21
93	Railways	1.92	0.13	0.13	0.00	0.08	2.25
94	Road transport	1.89	0.20	0.00	0.01	0.00	2.07
95	Sea and water transport	0.08	0.00	0.01	0.00	1.24	1.32
96	Air transport, etc.	0.43	0.08	0.01	0.00	0.69	1.21
97	Communication	2.12	0.48	1.11	0.00	0.33	4.04
98	Distributive trades	33.63	1.06	0.69	0.04	5.29	40.64
99	Insurance, banking, etc.	2.14	0.03	0.00	0.00	1.37	3.55
100	Property owning, etc.	1.34	0.12	0.09	0.00	0.00	1.55
101	Lodging and catering	11.02	0.50	0.00	0.00	1.75	13.27
102	Other services	6.98	3.73	2.39	0.01	1.31	14.41
110	Total output	256.79	41.49	127.31	33.88	512.26	975.90

Note: This table shows the total requirements of synthetic resins and plastics materials for each industry, induced by the final demand for the gross output of these industries. It is obtained by multiplying Row 29 of Table E of the *Input–Output Tables for the United Kingdom, 1974*, by the components of final demand for all industries which appear in Table D of the same input–output tables for 1974 (see Ref. 21). The sums along each row may not add up to round figures due to rounding errors. The total output figure of 979.90 is obtained by adding vertically downward along the total output column.

Source: Resources Policy, September 1984.

Table C2 Wholesale price index for plastics materials and semi-finished products in the United Kingdom (1975 = 100)

	1974	*1975*	*1976*	*1977*	*1978*
MLH 276.1 *Synthetic resins and plastics materials*	86.5	100.0	117.9	138.1	141.4
PVC	84.8	100.0	126.4	152.3	156.0
Polyethylene granules, compounds and powders	86.5	100.0	124.7	139.1	135.0
Styrene polymers & copolymers	88.2	100.0	118.3	134.7	134.1
Acrylics	81.5	100.0	113.2	125.2	129.2
Cellulosics and other plastics and modified natural resins	82.5	100.0	109.5	124.5	134.8
Semi-finished products	86.5	100.0	115.0	138.6	145.6
Order V chemicals and allied industries	82.1	100.0	114.8	133.1	143.7

Source: British Plastics Federation, annual report, 1980.

Table C3a Statistics on energy prices paid by the chemical industry: natural gas (US$ per GJ)

		1973	1974				1975				1976				1977				1978			
		July	*Jan*	*Apr*	*July*	*Oct*	*Jan*	*Apr*	*July*	*Oct*	*Jan*	*Apr*	*July*	*Oct*	*Jan*	*Apr*	*July*	*Oct*	*Jan*	*Apr*	*July*	*Oct*
Belgium	ni	0.69	0.61	0.68	0.78	0.88	1.37	1.41	1.31	1.38	1.76	1.83	1.78	1.97								
	i														1.79	1.88	1.95	2.04	2.20	2.24	2.16	2.31
Federal Republic of Germany	ni*	0.98	0.85	0.98	0.98	0.92	1.07	1.80	1.67	1.64	1.67	1.71	1.67	2.05	2.12	2.15	2.15	2.31	2.52	2.56	2.51	2.59
	ni†									2.15	2.21	1.96		2.37	2.43	2.43	2.50	2.84	2.90	2.94	2.88	2.98
France		1.00	1.34	1.32	1.43	1.43	1.91	1.96	1.89	1.82	1.84	1.97	1.94	1.87	1.85	1.97	1.99	2.19	2.50	2.57	2.61	2.72
Italy	ni	0.70	0.67	0.67	1.20	1.20	1.46	1.46	1.37	1.49	1.46	1.23	1.34	1.58	1.85	1.88	2.09	2.17	2.21	2.39	2.35	2.28
	i						1.80	2.08	1.94	1.70	1.40	1.49	1.96	1.93	2.03	2.22	2.21	2.11	2.08	2.25	2.22	2.17
Netherlands	ni	0.92	0.92	0.98	0.98	0.98	1.90	1.93	1.74	1.71	2.05	2.05	2.02	2.15	2.40	2.37	2.40	2.40	2.58	2.65	2.57	2.61
	i	0.76	0.79	0.82	0.82	0.82	1.61	1.61	1.45	1.42	1.86	1.83	1.80									
United Kingdom	ni	na					na	na	na	na	na	3.06	2.94	2.79	3.45	3.66	3.66	3.71	4.06	3.76	3.78	4.00
	i	0.95					2.56	2.59	2.29	2.17	2.17	2.06	2.00	1.88	2.46	2.65	2.57	2.61	2.86	2.67	2.68	2.84
Spain	i	1.00	1.03	0.99	1.02	1.02	1.62	1.63	1.56	1.52	1.83	1.62	1.61	1.62	2.08	2.08	2.05	1.68	2.07	2.08	2.12	2.30

ni non interruptible   i interruptible   * old contracts   † new contracts   *Source:* CEFIC.

Table C3b Statistics on energy prices paid by the chemical industry: natural gas (100 = price paid on 1 July 1973)

Indices		*1973*	*1974*				*1975*				*1976*				*1977*				*1978*			
		July	*Jan*	*Apr*	*July*	*Oct*	*Jan*	*Apr*	*July*	*Oct*	*Jan*	*Apr*	*July*	*Oct*	*Jan*	*Apr*	*July*	*Oct*	*Jan*	*Apr*	*July*	*Oct*
Belgium	ni	100	103	108	121	141	195	199	205	224	282	288	288	303								
Federal Republic of Germany	ni*	100	100	107	108	109	109	190	191	191	191	189	190	220	220	226	227	233	234	228	230	221
France		100	154	154	170	170	202	202	202	202	202	226	226	226	226	240	240	263	289	289	289	289
Italy	ni	100	100	100	189	189	217	217	217	238	269	269	269	315	402	408	454	463	480	475	468	454
Netherlands	ni	100	112	112	112	112	199	199	199	199	235	235	235	235	253	253	253	253	251	246	246	235
	i	100	113	113	113	113	198	198	198	198	253	253	253									
United Kingdom	i	100					287	287	287	287	287	301	301	301	382	413	401	401	405	389	389	389
Spain	i	100	100	100	100	100	156	156	156	156	188	188	188	188	244	244	244	244	284	285	285	285

ni non interruptible i interruptible * old contracts † new contracts Calculated on national currency basis

Source: CEFIC.

Table C4a Statistics on energy prices paid by the chemical industry: heavy fuel oil (US$ per GJ)

		1973	*1974*				*1975*				*1976*				*1977*				*1978*			
		July	*Jan*	*Apr*	*July*	*Oct*	*Jan*	*Apr*	*July*	*Oct*	*Jan*	*Apr*	*July*	*Oct*	*Jan*	*Apr*	*July*	*Oct*	*Jan*	*Apr*	*July*	*Oct*
Belgium	a	1.01	1.83	1.95	1.63	1.70	2.04	2.12/ 2.18	1.71/ 1.76	1.73/ 1.78	1.82/ 1.87	2.01/ 2.06	1.95/ 2.00	2.14/ 2.19	2.24/ 2.29	2.18/ 2.30	2.21/ 2.27	2.23/ 2.28	2.41/ 2.47	2.36/ 2.42	2.20/ 2.26	2.27/ 2.34
	b												1.75	1.89	2.22	1.96	2.03	2.08	2.14	2.10	2.01	2.02
Federal Republic of Germany	c	1.00	1.15	1.88	1.79	1.72	2.21	2.10	1.77	1.58	1.74	1.94	1.90	2.12	2.23	2.25	2.12	2.13	2.51	2.38	2.33	2.26
France	d	0.78	1.56	1.54	1.67	1.81	2.21	2.22	2.14	1.93	1.96	1.88	1.99	2.06	2.04	2.19	2.22	2.27	2.42	2.54	2.52	2.39
	e	0.65	1.36	1.34	1.48	1.61	2.00	2.00	1.79	1.72	1.75	1.68	1.80	1.82	1.81	1.95	1.97	2.03	2.17	2.28	2.25	2.12
Italy	f	(0.62)	(0.97)	(1.66)	(1.56)	(1.74)	(2.08)	(2.08)	(1.98)	(1.98)	1.55	1.49	1.86	1.95	1.96	2.16	2.17	2.17	2.19	2.34	2.29	2.22
	g	0.56	0.79	1.34	1.29	1.56	1.92	1.92	1.69	1.55	1.35	1.38	1.78	1.86	1.87	2.06	2.01	1.94	1.93	2.09	2.02	2.00
Netherlands	h	0.99	1.01	1.01	1.83	1.81	2.17	2.31	1.86	1.77	1.91	2.01	1.97	2.22	2.43	2.40	2.37	2.48				
	g	0.77	0.79	1.57	1.59	1.74													2.59	2.60	2.48	2.43
	j						2.05	2.11	1.76	1.67	1.79	1.96							2.48	2.44	2.33	2.31
	k												1.90	2.09	2.21	2.21	2.16	2.31				
United Kingdom	l	0.64					2.05	2.17	1.93	1.84	2.15	1.91	1.89	1.92	2.48	2.60	2.47	2.50				
	m																		2.73	2.63	2.47	2.59
Spain	n	0.70	0.72	1.33	1.37	1.37	1.40	1.76	2.11	2.06	2.06	1.82	1.81	1.82	1.80	1.98	1.95	1.83	2.06	2.08	2.11	2.29
	o	0.70	0.72	1.33	1.37	1.37	1.40	1.76	2.11	2.06	2.06	1.82	1.81	1.82	1.80	1.90	1.95	1.83	2.06	2.08	2.11	2.29

Source: CEFIC.

Table C4b Statistics on energy prices paid by the chemical industry: heavy fuel oil (100 = price at 1 July 1973)

		1973	1974				1975				1976				1977				1978			
		July	*Jan*	*Apr*	*July*	*Oct*	*Jan*	*Apr*	*July*	*Oct*	*Jan*	*Apr*	*July*	*Oct*	*Jan*	*Apr*	*July*	*Oct*	*Jan*	*Apr*	*July*	*Oct*
Belgium	a	100	211	211	172	186	199	204/	184/	193/	199/	217/	215/	224/	224/	222/	222/	222/	221/	207/	200/	194/
Federal Republic of Germany	c	100	133	204	197	197	223	212	196	182	194	212	211	222	226	231	213	212	228	208	208	188
France	d	100	232	232	254	275	302	295	276	276	276	276	298	321	321	344	344	352	359	367	358	326
	e	100	242	242	269	293	326	318	295	295	295	295	321	339	339	367	367	376	385	395	383	346
Italy	f	100	(167)	(287)	(287)	(320)	(360)	(360)	(365)	(365)	330	373	429	449	491	542	545	545	549	540	527	510
	g	100	150	261	261	312	366	366	342	314	316	382	452	471	503	551	538	518	516	511	496	492
Netherlands	h	100	112	191	191	193	207	217	194	190	200	211	211	224	234	235	230	239	232	221	218	201
	g	100	112	212	212	237													284	266	261	244
United Kingdom	l	100					343	358	358	358	423	412	423	456	580	601	571	571	571	566	519	519
Spain	n	100	100	194	194	194	194	242	303	303	303	303	303	303	303	333	333	382	409	409	409	409

Calculated on national currency basis *Source:* CEFIC.

Table C5 Gross domestic fixed capital formation: in new buildings and works*; by sector: United Kingdom (1970–1980)

	£ million										
	1970	*1971*	*1972*	*1973*	*1974*	*1975*	*1976*	*1977*	*1978*	*1979*	*1980*
(a) At current prices											
Private sector	1,261	1,535	1,865	2,164	2,523	3,204	3,720	4,246	5,166	6,411	8,260
General government	1,510	1,598	1,748	2,453	2,717	2,858	2,936	2,389	2,103	2,267	2,473
Public corporation	424	443	370	414	924	1,498	1,638	1,728	1,882	2,095	2,450
All sectors†	3,195	3,576	3,983	5,031	6,164	7,560	8,294	8,363	9,151	10,773	13,183
(b) At constant 1975 prices											
Private sector	3,037	3,381	3,527	3,306	2,942	3,204	3,444	3,586	3,890	3,846	3,960
General government	3,455	3,338	3,253	3,853	3,373	2,858	2,633	2,046	1,662	1,524	1,267
Public corporations	1,046	994	757	693	1,154	1,498	1,468	1,468	1,457	1,357	1,240
All sectors†	7,655	7,781	7,607	7,852	7,469	7,560	7,545	7,110	7,009	6,727	6,467

* Excluding dwellings but including purchases less sales of land and existing buildings.

† The value of land and existing buildings which pass from one sector to another are netted out over all sectors. The total figures relate to the costs of transferring the ownership of such assets.

Source: Housing and Construction Statistics, HMSO, London, 1982.

Table C6 Gross domestic fixed capital formation in housing: United Kingom (1970–1980)

	£ million										
	1970	*1971*	*1972*	*1973*	*1974*	*1975*	*1976*	*1977*	*1978*	*1979*	*1980*
(a) At current prices											
Public sector	802	798	819	1,020	1,481	1,967	2,312	2,225	2,208	2,320	2,551
Private sector	841	1,100	1,435	1,666	1,706	2,182	2,414	2,466	3,169	3,250	3,482
All housing	1,643	1,898	2,254	2,686	3,187	4,149	4,726	4,691	5,377	5,570	6,033
All housing as percentage of gross domestic product at factor cost	3.77	3.84	4.07	4.17	4.27	4.39	4.24	3.70	3.70	3.35	3.12
At 1975 prices * †											
Public sector	1,879	1,733	1,609	1,703	1,822	1,967	2,085	1,869	1,740	1,581	1,418
Private sector	2,047	2,440	2,745	2,449	2,004	2,182	2,180	2,039	2,327	1,975	1,711
All housing	3,850	4,091	4,309	4,152	3,826	4,149	4,265	3,908	4,067	3,556	3,129
All housing as percentage of gross domestic product at factor cost	4.50	4.67	4.85	4.34	4.03	4.39	4.35	3.94	3.99	3.45	3.09

* For the years prior to 1973, totals may not equal the sum of their components due to method used in rebasing to 1975 prices.
† The 1975 based price indices have been revised.

Source: Housing and Construction Statistics, HMSO, London, 1982.

Table C7 Material prices' indices for the United Kingdom construction industry, 1971–1981 (1975 = 100)

	Metals					Other materials			
	Steel								
Year	*Heavy rolled products – 80 mm and over*	*Tubes*	*Copper tubes*	*Iron – cast and spun pipes and fittings*	*Plumbers' brassware*	*Window frames and doors*	*Glass – sheet and plate*	*Paint*	*Plastic building materials*
1971	46	46	79	46	63		66	55	54
1972	49	49	79	49	66		69	59	57
1973	54	53	110	53	77		71	63	61
1974	76	75	134	71	93	80	79	78	86
1975	100	100	100	100	100	100	100	100	100
1976	118	131	132	124	123	125	122	110	118
1977	143	144	141	138	142	151	150	126	142
1978	161	164	143	153	155	167	163	141	155
1979	169	178	194	179	188	193	176	160	188
1980	176	189	214	208	234	233	204	192	224
1981	177	196	208	222	249	254	223	205	232

Source: *Housing and Construction Statistics: 1971–1981*, HMSO, London, 1982, p. 53.

Table C8 Imports and exports of electrical wires and plastics materials for construction use: the United Kingdom, 1975–1981

	at 1975 prices, £000s					
	1975	*1976*	*1977*	*1978*	*1979*	*1980*
Imports						
Electrical wires	5,547	7,174	8,915	10,974	15,476	19,925
Plastic	25,185	43,339	54,086	77,472	85,631	88,493
Exports						
Electrical wires	38,217	44,487	52,064	46,282	43,740	49,242
Plastic	49,108	71,746	92,522	116,632	114,809	128,829

Source: Housing and Construction Statistics: 1971–1981, HMSO, London, 1982, p. 59.

APPENDIX D:

STATISTICAL DATA FOR CHAPTER 4

Table D1 Existing petrochemical facilities and planned capacities in OPEC member countries

Complex site	*Product*	*Capacity (tonnes/year)*	*Comments*
ALGERIA		*Existing petrochemical capacities*	
Arzew	ammonia	330,000	start-up 1970
	urea	130,000	1970
	methanol	100,000	1976
Skikda	ethylene	120,000	start-up 1978
	VCM	40,000	1980
	PVC	35,000	1980
	LDPE	48,000	1981
Arzew	ammonia	330,000	start-up 1981
Annaba	ammonia		start-up 1982
		Projects planned or under construction	
Skikda	benzene	90,000	start-up 1983
	toluene	5,000	1983
	xylenes	157,000	1983
	PVC	20,000	1983
ECUADOR		*Existing petrochemical facilities*	
Tintessa, Quito	polyester resins	2,400	start-up 1976

Table D1—*cont.*

Complex site	*Product*	*Capacity (tonnes/year)*	*Comments*
		Projects planned or under construction	
President Jaime Roldos	ammonia	330,000	start-up 1983
Guayaquil-port	urea	561,000	1985. Exports to Andean market
	ethylene	140,000	
	HDPE	63,000	start-up 1987 50% for domestic market
	polypropylene	72,000	start-up 1987 30% for domestic market
President Jaime Roldos	PVC	40,000	1987
Guayaquil-port	butadiene	25,000	1987
GABON		*Existing petrochemical capacities*	
	PVC	20,000	start-up 1980
	PVAC & alkyds	15,000	based on imported basic materials
	urea formol	10,000	
INDONESIA		*Existing petrochemical capacities*	
Plaju (South Sumatra)	polypropylene	20,000	start-up 1973 LPG feedstock
	ammonia	1,050,000	four facilities on-stream

Table D1—*cont.*

Complex site	*Product*	*Capacity (tonnes/year)*	*Comments*
Cikampek (West Java)	ammonia	330,000	start-up 1979/80
	urea	560,000	1979/80
Kaltim I & II (North Sumatra)	ammonia	495,000	start-up 1982
	urea	560,000	1982
	ammonia	330,000	ASEAN venture
	urea	570,000	60% Indonesian
		Projects planned or under construction	
Arun (North Sumatra)	ethylene	330,000	start-up 1985/86
	ethylene glycol	70,000	ethane gas
	LDPE	150,000	feedstock
	HDPE	100,000	
	VCM	110,000	
Plaju	benzene	374,000	start-up 1985
	xylenes	180,000	
	dimethylterphthalate	80,000	
Bunyu (Java)	methanol	330,000	start-up 1983
IRAN		*Existing and planned petrochemical capacities*	
Shiraz	ammonia	40,000	start-up 1963
	urea	50,000	locally consumed

Table D1—*cont.*

Complex site	*Product*	*Capacity (tonnes/year)*	*Comments*
Abadan	ethylene	26,000	
	PVC	60,000	
	VCM	63,000	
Bandar Khomeini	ammonia	330,000	start-up 1970 and 1978
	urea	164,000	
	ethylene	300,000	
	LDPE	100,000	
	HDPE	60,000	
	propylene	110,000	
	polypropylene	50,000	
	PVC	150,000	start-up 1983
	DMT	70,000	
	SBR	40,000	
	xylenes	140,000	
	aromatics	500,000	start-up 1982
IRAQ		*Existing and planned petrochemical capacities*	
Basrah	ammonia	66,000	start-up 1971
	urea	53,000	
	ammonia	264,000	start-up 1978
	urea	429,000	
Khor-Alzubair	ammonia	660,000	start-up 1980
	urea	1,056,000	

Table D1—*cont.*

Complex site	*Product*	*Capacity (tonnes/year)*	*Comments*
Basrah	PVC	60,000	start-up 1980
	ethylene	150,000	start-up 1984
	LDPE	60,000	
	HDPE	30,000	
KUWAIT		*Existing petrochemical capacities*	
Shuaiba	ammonia	660,000	
	urea	792,000	
		Projects planned or under construction	
Shuaiba	ethylene	330,000	start-up 1984
	LDPE	130,000	
	ethylene glycol	135,000	
	styrene	340,000	
	benzene	283,000	
	xylenes	146,000	
	ammonia	330,000	
Sitra	ammonia	330,000	start-up 1984
	methanol	330,000	kpc, SABIC and Bahrain NOC. Joint venture

Table D1—*cont.*

Complex site	*Product*	*Capacity (tonnes/year)*	*Comments*
LIBYA	*Existing petrochemical facilities*		
Marsa El Brega	methanol	330,000	start-up 1978
	ammonia	330,000	NG feedstock
	urea	330,000	start-up 1980
	ammonia	330,000	start-up 1982
Ras Lanuf	ethylene	330,000	start-up 1982
	LDPE	50,000	
	HDPE	50,000	
	polypropylene	50,000	
	ethylene glycol	52,000	
	butadiene	58,000	
Marsa El Brega	methanol	330,000	start-up 1983
	urea	577,500	start-up 1984
	ammonia	660,000	under study
	urea	990,000	under study
NIGERIA	*Projects planned or under construction*		
Warri	polypropylene	35,000	start-up 1983 LPG feedstock
Kaduna	benzene	25,000	start-up 1983
	ethylene	300,000	under study
	LDPE	110,000	to start-up in 1987

Table D1—*cont.*

Complex site	*Product*	*Capacity (tonnes/year)*	*Comments*
	HDPE	70,000	
	polypropylene	60,000	
	PVC	140,000	
	ethylene glycol	35,000	
	polypropylene/oxide polyols	40,000	
QATAR		*Existing and planned petrochemical facilities*	
Umm Said	ammonia	594,000	mainly for export
	urea	660,000	
	ethylene	280,000	start-up 1979
	LDPE	140,000	for export
	HDPE	70,000	under construction
Dunkirk	LDPE	100,000	start-up 1982 joint venture with CdF-Chimie
UNITED ARAB EMIRATES		*Planned or under construction capacities*	
	ammonia	330,000	under construction
		At present the UAE has no existing petrochemical industry.	

Table D1—*cont.*

Complex site	*Product*	*Capacity (tonnes/year)*	*Comments*
SAUDI ARABIA		*Existing petrochemical facilities*	
Dammam	ammonia	200,000	on stream
	urea	300,000	
		Projects planned or under construction	
Dammam	urea	300,000	start-up 1982
	ammonia	330,000	
Al Jubayl	methanol	600,000	start-up 1983
	urea	500,000	
	methanol	600,000	start-up 1984
Yanbu	ethylene	450,000	start-up 1984
	LDPE	200,000	
	HDPE	90,000	
	ethylene glycol	200,000	
Al Jubayl	ethylene	650,000	start-up 1985
	ethanol	280,000	
	ethylene dichloride	454,000	
	styrene	295,000	
	ethylene	500,000	
	HDPE and LDPE	180,000	
	ethylene glycol	300,000	
	LDPE	260,000	
	LDPE	130,000	
	ethylene glycol	300,000	

Table D1—*cont.*

Complex site	*Product*	*Capacity (tonnes/year)*	*Comments*
VENEZUELA		*Existing petrochemical facilities*	
Zulia	polystyrene	36,000	start-up 1973
	polypropylene	18,000	Andean market
	PVC	45,000	
Moron	ammonia	198,000	start-up 1983
	urea	248,000	
	benzene	18,000	start-up 1969 South American market
Zulia	ammonia	594,000	start-up 1974
	urea	792,000	Andean market and Europe
	ethylene	150,000	start-up 1975
	propylene	35,000	LPG feedstock
	LDPE	50,000	start-up 1976 Andean market
		Projects planned or under construction	
Zulia	LDPE	60,000	start-up 1983 expansion of existing plant
	HDPE	60,000	start-up 1983

Source: OPEC, *Annual Report*, 1981.

Table D2 Typical composition of associated natural gases (mol. per cent)

	Venezuela	*Saudi Arabia*	*Nigeria*	*Iran*	*No. Sea Ekofisk*	*Kuwait*	*Abu Dhabi*	*Libyan Arab Jamahiriya*
Nitrogen	0.4	0.3	0.1	—	0.5	—	0.2	—
Carbon dioxide	0.2	11.1	1.0	0.3	1.6	1.6	4.4	—
Hydrogen sulphide	—	2.7	—	—	—	0.1	1.7	—
Methane	85	48.1	83.6	74.9	85.9	78.2	76.4	66.8
Ethane	8	18.6	6.8	13.0	8.1	12.6	8.1	19.4
Propane	4.4	11.7	4.5	7.2	2.7	5.1	4.7	9.1
Butane	2	4.6	2.6	3.1	0.9	0.6	2.6	3.5
Natural gasoline	—	2.9	1.4	1.5	0.3	0.8	1.8	1.2
Helium	—	—	—	—	—	—	—	—

Source: UNIDO and GOIC PC.24, November 1981.

Table D3 Installed cost for petrochemical plants in 1980 ($/tonne/year at 100 per cent load factor)

Location		*US Gulf Coast*	*Federal Republic of Germany*	*Japan*	*Indonesia*	*Mexico*	*Qatar*
(Location factor)		*(1.00)*	*(1.15)*	*(0.90)*	*(2.1)*	*(1.25)*	*(1.5)*
Product	*Capacity range 1,000 tonne/year*	*Installed cost range $/tonne/year*	*Installed cost range $/tonne/year*	*Installed cost range $/tonne/year*	*Installed cost range $/tonne/year*	*Installed cost range $/tonne/year*	*Installed cost range $/tonne/year*
Ammonia							
from methane	300–590	277–313	318–360	249–282	281–657	346–391	415–469
from naphtha	300–590	317–356	364–409	285–320	665–747	396–444	475–533
DMT	75–300	883–1,181	1,015–1,358	795–1,063	1,854–2,480	1,104–1,477	1,324–1,772
Ethyl benzene	250–780	77–112	88–129	69–101	161–235	96–140	115–168
Ethylene-propylene*	225–680	611–802	703–922	550–722	1,284–1,684	764–1,002	917–1,202
Ethylene-propylene-butadiene-benzene†	225–680	787–1,025	905–1,179	708–923	1,653–2,153	984–1,282	1,181–1,538
Ethylene glycol	90–360	153–234	176–270	137–211	321–492	191–293	229–352
Ethylene oxide	67–270	701–1,006	806,1,157	137–905	1,472–2,112	876–1,257	1,052–1,509
HDPE	50–200	478–640	550–736	431–576	1,004–1,344	598–800	718–960
LDPE	50–200	692–1,000	796–1,150	623–900	1,453–2,100	865–1,250	1,038–1,500
LLDPE	50–200	461–634	530–729	415–571	968–1,331	576–792	691–951

Table D3—*cont.*

Location		*US Gulf Coast*	*Federal Republic of Germany*	*Japan*	*Indonesia*	*Mexico*	*Qatar*
(Location factor)		*(1.00)*	*(1.15)*	*(0.90)*	*(2.1)*	*(1.25)*	*(1.5)*
Product	*Capacity range 1,000 tonne/year*	*Installed cost range $/tonne/year*	*Installed cost range $/tonne/year*	*Installed cost range $/tonne/year*	*Installed cost range $/tonne/year*	*Installed cost range $/tonne/year*	*Installed cost range $/tonne/year*
Methanol							
from methane	160–640	206–287	237–330	185–258	432–602	257–358	304–430
from naphtha	160–640	225–125	258–373	202–292	472–682	281–406	337–487
Polyethylene terephthalate (PET)							
from DMT	22–90	828–1,178	852–1,354	745–1,060	1,738–2,473	1,034–1,472	1,242–1,767
from TPA	25–100	694–1,116	798–1,283	625–1,004	1,457–2,344	867–1,395	1,041–1,674
Polypropylene	45–180	799–1,013	919–1,165	719–912	1,679–2,128	999–1,267	1,199–1,520
Polystyrene	45–180	352–486	404–560	316–438	738–1,022	439–608	527–730
PVC	150–500	645–998	741–1,148	580–898	1,354–2,096	806–1,247	967–1,497
SBR	35–140	856–1,331	949–1,531	771–1,198	1,798–2,796	1,070–1,664	1,285–1,997
Styrene	225–680	215–282	247–324	193–254	451–593	268–353	322–423
Terephthalic acid (TPA)	75–300	863–1,117	993–1,285	777–1,005	1,813–2,346	1,079–1,397	1,295–1,676
Urea	245–860	91–136	104–156	82–122	190–295	113–170	136–204
VCM	180–730	311–414	257–476	280–372	653–869	388–517	466–621

* Cost per tonne ethylene-propane feedstock. † Cost per tonne ethylene from naphtha feedstock.

Source: UNIDO, 1981, 236/2

Table D4 Freight costs for shipping petrochemicals to industrialised country markets in 1980 (1980 dollars)

Producer/exporter Product	*Destination market*			
	Japan	*Northern Europe*	*Southern Europe*	*United States*
	$/ton	*$/ton*	*$/ton*	*$/ton*
Qatar				
LPG	41.1	45.5	34.4	59.0
Ammonia	35.4	39.1	29.6	50.1
Ethylene	43.2	47.8	36.1	61.2
Propylene	40.3	44.6	48.2	78.4
Methanol	19.2	21.8	16.6	29.0
Urea (bulk)	39.2	41.6	33.1	54.2
Urea (bagged)	67.0	69.2	58.4	84.6
PVC (bagged)	85.1	88.2	72.5	110.7
LDPE (bagged)	92.3	95.7	78.7	120.0
Mexico				
LPG	59.1	33.3	56.1	9.0
Ammonia	50.8	28.6	48.2	7.7
Ethylene	62.1	35.0	58.9	9.4
Propylene	67.4	38.0	63.9	10.3
Methanol	27.9	15.6	16.6	4.0
Urea (bulk)	54.1	31.9	34.8	11.5
Urea (bagged)	85.1	58.0	66.1	31.7
PVC (bagged)	109.9	72.0	83.7	34.4
LDPE (bagged)	119.3	78.1	90.8	37.3
Indonesia				
LPG	24.3	53.1	42.3	68.8
Ammonia	20.9	45.7	36.4	59.1
Ethylene	25.5	55.7	44.4	72.2
Propylene	27.7	60.5	48.2	78.4
Methanol	11.4	25.2	20.2	35.5
Urea (bulk)	24.9	48.1	39.6	61.1
Urea (bagged)	49.2	77.2	66.6	93.2
PVC (bagged)	59.4	99.8	84.4	123.0
LDPE (bagged)	64.4	108.2	91.6	133.5

Source: UNIDO, 1981, 336/2.

Table D5 Post-MTN average tariff rates for selected petrochemical products (*ad valorum* or *ad valorum* equivalent)

Product	*EEC*	*Japan*	*USA*	*Austria*	*Australia*	*Canada*	*Finland*	*New Zealand*	*Norway*	*Sweden*	*Switzer-land*
Basic feedstocks											
Ethylene	6.3	6	F	F	F	F	F	F	F	6.5	—
Propylene	6.3	6	F	F	F	F	F	F	F	6.5	—
Butadiene	6.3	6	F	F	F	F	F	F	F	6.5	—
Benzene	F	4	F	10	F	F	2.2	F	F	F	—
Xylene-para-ortho	F	2	F	10	F	F	F	F	F	F	—
Ammonia	11	4	3	22	F	F	F	13	F	9	6
Methanol	13	5	18	15	19	10	F	F	7.7	9	—
Intermediates											
Polyethylene-LD	12.5	11	12.5	21	22.5	9.5	F	F	10	9	2
-HD	12.5	11	12.5	21	37.5	9.5	F	F	10	9	2
Polystyrene	12.5	14	17	21	22.5	9.5	F	F	20	9	2
Polypropylene	12.5	22	12.5	21	22.5	9.5	F	F	20	9	2
PVC	12.5	6	10.1	18	22.5	9.5	F	F	20	9	5
SBR	3	F	F	F	37.5	F	F	F	7.7	F	—
Fibres											
Polyester	7.5	10	9	F	F	8.5	F	F	2	3.2	6
Polyamide	7.5	10	5	F	7.5	8.5	F	F	2	3.2	6

F duty-free — less than 1% *ad valorum* *Source:* UNIDO, 1981, 336/3, p. 186.

The above tariff rates refer to the protocol of the general agreement on tariffs and trade, 30th June 1979, Geneva and the protocol supplementary to the Geneva agreement, 22nd November 1979.

Table D6a Production capacity of thirty developing countries in three basic petrochemicals (olefins) (thousand metric tonnes)

	Ethylene			*Propylene*			*Butadiene*		
	1979	*1984*	*1987*	*1979*	*1984*	*1987*	*1979*	*1984*	*1987*
Africa									
Nigeria	—	—	280	—	35	35	—	—	—
North Africa									
Algeria	120	120	120	—	—	—	—	—	—
Egypt	—	—	140	—	—	—	—	—	—
Libya	—	330	330	—	50	50	—	60	60
Morocco	—	—	—	—	—	—	—	—	—
W. Asia									
Bahrain	—	—	—	—	—	—	—	—	—
Iraq	30	160	160	—	—	—	—	—	—
Kuwait	—	—	300	—	—	—	—	—	—
Qatar	—	280	280	—	—	—	—	—	—
Saudi Arabia	—	—	1,600	—	—	—	—	—	—
Turkey	60	360	260	40	100	100	30	30	30
United Arab Emirates	—	—	—	—	—	—	—	—	—
Asia									
India	240	240	920	120	120	220	50	50	70
Indonesia	—	—	350	—	—	—	—	—	—
Iran	30	30	300	15	15	125	—	—	25

Table D6a—*cont.*

	Ethylene			*Propylene*			*Butadiene*		
	1979	*1984*	*1987*	*1979*	*1984*	*1987*	*1979*	*1984*	*1987*
Malaysia	—	—	—	—	—	—	—	—	—
Pakistan	—	—	100	—	—	—	—	—	—
Philippines	—	—	250	—	—	—	—	—	—
Rep. of Korea	150	850	1,200	80	450	630	25	125	175
Singapore	—	300	300	—	165	165	—	—	—
Thailand	—	—	150	—	—	—	—	—	—
Other Asia	570	920	920	290	410	410	80	120	120
China	540	950	1,810	230	410	950	100	130	220
Latin America									
Argentina	170	250	550	80	100	240	40	40	120
Bolivia	—	—	160	—	—	80	—	—	—
Brazil	740	1,220	1,220	410	650	650	170	240	240
Chile	60	180	180	—	—	—	—	—	—
Columbia	20	120	120	10	10	10	—	—	—
Ecuador	—	—	100	—	—	—	—	—	—
Mexico	440	1,440	1,940	150	450	450	150	150	250
Peru	—	—	250	—	—	150	—	—	70
Venezuela	150	150	500	90	90	90	—	—	—

Table D6b Production capacity of thirty developing countries in three basic petrochemicals (aromatics and methanol) (thousand metric tonnes)

	*Xylenes**			*Benzene*			*Methanol*		
	1979	*1984*	*1987*	*1979*	*1984*	*1987*	*1979*	*1984*	*1987*
Africa									
Nigeria	—	—	—	—	20	20	—	—	—
North Africa									
Algeria	—	40	40	—	—	—	110	110	110
Egypt	—	—	—	—	—	—	—	—	—
Libya	—	—	—	—	—	—	330	330	330
W. Asia									
Bahrain	—	—	—	—	—	—	—	330	330
Iraq	—	—	—	—	—	—	—	330	330
Kuwait	—	—	140	—	—	280	—	—	—
Qatar	—	—	—	—	—	—	—	—	—
Saudi Arabia	—	—	—	—	—	—	—	—	1,400
Turkey	—	200	200	—	150	150	—	100	100
United Arab Emirates	—	—	—	—	—	—	—	—	—
Asia									
India	40	100	160	150	210	310	—	60	60
Indonesia	—	—	240	—	—	370	—	—	330
Iran	—	—	120	—	—	350	—	—	100

Table D6b—*cont.*

	*Xylenes**			*Benzene*			*Methanol*		
	1979	*1984*	*1987*	*1979*	*1984*	*1987*	*1979*	*1984*	*1987*
Malaysia	—	—	—	—	—	—	—	—	330
Pakistan	40	40	40	—	—	—	—	—	—
Philippines	—	—	—	—	—	—	—	—	—
Rep. of Korea	50	400	400	110	250	250	390	390	390
Singapore	—	—	—	—	—	—	—	—	—
Thailand	—	—	—	—	—	—	—	—	—
Other Asia	260	400	400	200	340	440	120	190	190
China	30	210	400	400	500	800	260	400	800
Latin America									
Argentina	65	65	65	230	230	290	40	40	40
Bolivia	—	—	—	—	—	100	—	—	—
Brazil	160	230	230	270	390	390	140	140	140
Chile	—	—	—	—	—	—	—	—	—
Columbia	60	60	210	40	40	90	—	—	—
Ecuador	—	—	70	—	—	140	—	—	—
Mexico	110	410	710	120	720	720	180	1,000	1,820
Peru	—	—	20	—	—	120	—	—	—
Venezuela	—	—	50	—	—	100	—	—	330

* Para-Xylene and Ortho-Xylene

Source: UNIDO, 1981, 336/3.

Table D7 Production capacity of thirty developing countries in five thermoplastics (thousand metric tonnes)

	LDPE			*HDPE*			*Polypropylene*		
	1979	*1984*	*1987*	*1979*	*1984*	*1987*	*1979*	*1984*	*1987*
Africa									
Nigeria	—	—	110	—	—	70	—	35	35
North Africa									
Algeria	48	48	48	—	—	—	—	—	—
Egypt	9	90	90	—	40	40	—	—	—
Libya	50	50	50	50	50	50	70	70	70
Morocco	—	—	—	—	—	—	—	—	—
Western Asia									
Bahrain	—	—	—	—	—	—	—	—	—
Iraq	—	60	60	—	30	30	—	—	—
Kuwait	—	—	130	—	—	—	—	—	—
Qatar	—	140	140	—	70	70	—	—	—
Saudi Arabia	—	—	680	—	—	160	—	—	—
Turkey	25	175	175	—	40	40	—	60	60
United Arab Emirates	—	—	—	—	—	—	—	—	—
Asia									
India	112	112	N.E.	30	30	N.E.	30	30	N.E.
Indonesia	—	—	180	—	—	60	37	37	—
Iran	—	—	100	—	—	60	—	—	50

Table D7—*cont.*

	LDPE			*HDPE*			*Polypropylene*		
	1979	*1984*	*1987*	*1979*	*1984*	*1987*	*1979*	*1984*	*1987*
Malaysia	—	—	—	—	—	—	—	—	—
Pakistan	5	5	65	—	—	—	—	—	—
Philippines	—	—	100	—	—	35	—	—	—
Rep. of Korea	70	320	N.E.	70	140	N.E.	125	205	N.E.
Singapore	—	120	120	—	80	80	—	100	100
Thailand	—	74	74	—	—	—	—	—	—
Other Asia	215	215	N.E.	50	170	N.E.	50	185	N.E.
China	280	340	860	35	35	35	120	200	440
Latin America									
Argentina	33	228	N.E.	—	30	N.E.	—	40	N.E.
Bolivia	—	—	40	—	—	95	—	—	40
Brazil	320	570	N.E.	110	170	N.E.	100	150	N.E.
Chile	36	36	36	—	—	—	—	—	—
Columbia	20	60	60	—	—	—	—	—	—
Ecuador	—	—	60	—	—	35	—	—	40
Mexico	100	340	N.E.	100	200	N.E.	—	100	N.E.
Peru	—	—	90	—	—	—	—	—	—
Venezuela	56	110	110	—	60	60	—	—	—

N.E. no estimate

Table D7—*cont.*

	PVC			*Polystyrene*		
	1979	*1984*	*1987*	*1979*	*1984*	*1987*
Africa						
Nigeria	—	—	145	—	—	—
North Africa						
Algeria	35	35	35	—	—	—
Egypt	—	80	80	—	—	—
Libya	—	80	80	—	—	—
Morocco	25	—	—	—	—	—
Western Asia						
Bahrain	—	—	—	—	—	—
Iraq	—	60	60	—	—	—
Kuwait	—	—	—	—	—	—
Qatar	—	—	—	—	—	—
Saudi Arabia	—	—	—	—	—	—
Turkey	52	152	152	15	N.E.	N.E.
United Arab Emirates	—	—	—	—	—	—
Asia						
India	132	187	N.E.	24	24	N.E.
Indonesia	40	40	150	—	—	—
Iran	—	—	150	—	—	—

Table D7—*cont.*

	PVC			*Polystyrene*		
	1979	*1984*	*1987*	*1979*	*1984*	*1987*
Malaysia	25	25	—	7	—	—
Pakistan	5	5	55	—	—	—
Philippines	50	50	N.E.	13	13	N.E.
Rep. of Korea	200	300	N.E.	50	200	N.E.
Singapore	—	—	—	—	—	—
Thailand	20	50	N.E.	15	23	N.E.
Other Asia	400	1,000	N.E.	80	120	N.E.
China	400	800	800	20	20	20
Latin America						
Argentina	53	147	N.E.	57	57	N.E.
Bolivia	—	—	—	—	—	—
Brazil	311	511	N.E.	185	200	N.E.
Chile	15	N.E.	N.E.	4	N.E.	N.E.
Columbia	—	30	30	—	10	10
Ecuador	—	—	20	—	—	10
Mexico	120	260	N.E.	98	148	N.E.
Peru	—	—	60	—	—	36
Venezuela	45	45	90	36	36	54

N.E. no estimate

ABBREVIATIONS

ABS	Acrylonitrile – Butadiene – Styrene
AP	Amino plastics
ASEAN	Association of South East Asian Nations
ARAMCO	Arabian American Oil Company
BP	British Petroleum
BTX	Benzene –Toluene – Xylene
CdF Chimie	Carbonage de France –Chimie
CEFIC	Conseil Européen des Fédérations de l'Industrie Chimique
CFP	Compagnie Française des Pétroles
CMEA	Council for Mutual Economic Assistance (COMECON)
DMT	Di-methyl terephthalate
EDC	Ethylene di-chloride
EEC	European Economic Community
ENI	Ente Nazionale Idrocarburi
FCC	Fluid Catalytic Crackers
GDP	Gross Domestic Product
GOIC	Gulf Organisation for Industrial Consulting
HDPE	High density polyethylene
ICI	Imperial Chemical Industries
IEA	International Energy Agency
I/O	Input–output
IPCs	Independent petroleum-chemical companies
LDPE	Low density polyethylene
LPG	Liquefied petroleum gas
MCs	Major chemical companies
MPs	Major petroleum companies' chemical subsidiaries
NEDO	National Economic Development Office
NICs	Newly industrialised countries
NIOC	National Iranian Oil Company

OECD	Organization of Economic Co-operation and Development
OPDC	Oil-producing developing countries
OPEC	Organization of Petroleum Exporting Countries
O-Xylene	Ortho-xylene
PE	Polyethylene
PET	Polyethylene terephthalate
PP	Polypropylene
PS	Polystyrene
PVAC	Polyvinyl acetate
PVC	Polyvinyl chloride
P-Xylene	Para-xylene
ROI	Return on investment
$	US dollar
SABIC	Saudi Basic Industries Corporation
SBR	Styrene – Butadiene – Rubber
TOE	Ton of oil equivalent
TPA	Terephthalic acid
UK	United Kingdom
UN	United Nations
UNIDO	United National Industrial Development Organisation
USA	United States of America
USSR	Union of Soviet Socialist Republics
VCM	Vinyl chloride monomer

INDEX

Note: this index does not include personal or organisation names, or passing references to countries.